Lichens of California

Guides: 54

Lichens of California

Mason E. Hale, Jr.,
and Mariette Cole

UNIVERSITY OF CALIFORNIA PRESS
Berkeley Los Angeles London

University of California Press
Berkeley and Los Angeles, California

University of California Press, Ltd.
London, England

© 1988 by
The Regents of the University of California

Printed in the United States of America
1 2 3 4 5 6 7 8 9

Library of Congress Cataloging-in-Publication Data

Hale, Mason E.
  Lichens of California / by Mason E. Hale, Jr., and Mariette Cole.
    p.   cm. — (California natural history guides ; 54)
  Bibliography: p.
  Includes index.
  ISBN 0-520-05712-0 (alk. paper).
  ISBN 0-520-05713-9 (pbk. : alk. paper)
  1. Lichens—California.   2. Lichens—California—Identification.
  I. Cole, Mariette.   II. Title.   III. Series.
  QK587.5.C2H35   1988
  589.1'09794—dc19   87-18456

FIGURE CREDITS: Figs. 1, 5, 14, 64 (Hale, 1961). Fig. 2 (Ornduff, 1974). Figs. 6, 16, 17, 22b, 30, 50 (drawings by J. Schroeder). Figs. 18, 28 (drawings by J. Putnam). Figs. 53, 56, 61 (drawings by N. Halliday). *Cover illustration: Letharia vulpina*

Contents

Preface

Lichens are some of the most conspicuous and colorful plants in California. Nearly 1000 species have been found in the state, occurring almost everywhere on rocks and trees. The trunks of giant Sequoias are sometimes completely covered with the chartreuse-yellow Wolf Moss (*Letharia vulpina*), and the Valley Oaks are draped with Lace Lichen (*Ramalina menziesii*) and old man's beard (*Usnea* sp.).

Why is it, then, that most naturalists probably know less about lichens than any other plant group? True, lichens have little direct economic importance to us and are barely if at all mentioned in botany courses. Worse yet, there are almost no useful field guides—and none in the color so necessary to illustrate their brilliant orange and yellow hues. Where guides are available, they are often difficult for beginners because of the unfamiliar terminology and technical descriptions. And, finally, very few lichens have common names in English; they are known only by Latin scientific names.

This guide provides a useful introduction to lichens by removing some of the obstacles to appreciating this fascinating, hardy form of plant life. It includes full descriptions of 196 species and mentions 125 others—the majority of the foliose and fruticose species and examples of many crustose groups.

We are most grateful to the following colleagues who collected specimens, reviewed data, or helped in other ways: Dr. T. Ahti, Doris Baltzo, Charis Bratt, Dr. William Jordan, Beth Kantrud, Elmer Schmidt, Sue Sweet, Dr. Harry Thiers, Dr. Clifford Wetmore, and the late Dr. Frank McWhorter.

INTRODUCTION

What Are Lichens?

What are these arcane creatures called lichens? The word *lichen* comes from the Greek *leichen,* meaning a tree moss. Lichen is also a medical term used to describe various skin diseases in older people. Linnaeus, the great Swedish botanist of the 1700s, was the first to classify lichens, describing 80 species in the genus *Lichen.* He considered them to be algae, little more than trash of the plant kingdom.

For the next hundred years lichens were treated as a separate group of plants, coordinate with fungi and algae. Then, in 1867, the Swiss lichenologist Schwendener made a revolutionary discovery: Lichens are not a different kind of plant at all. They are a combination of two entirely different organisms, fungus and alga, growing together in a symbiotic or mutually beneficial union.

How can a lichen behave like an independent organism, such as a fern or moss, when by definition it is not? The main reason is the unique product of the symbiotic relationship. A moldlike fungus captures and entraps microscopic green algae—similar to those in green mats on damp trees—and in the process forms a new plant body which has no resemblance to either a fungus or an alga. This plant body, technically called a thallus, is totally self-sufficient; that is, the chlorophyll-containing algae utilize sunlight to produce organic foodstuffs that are absorbed by and sustain the fungus. The fungus in turn selectively parasitizes but does not kill off its generous host. The algae seem to benefit by protection of the fungus within

1

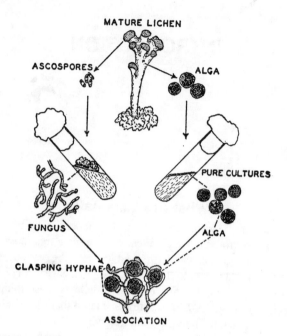

FIG. 1. Experimental steps in isolating and reconstituting the fungal and algal components of a lichen.

the thallus. The two components can be separated in the laboratory and experimentally recombined to form a "new" lichen, at least in a few species (fig. 1).

Lichens have very simple physiological demands. Mineral nutrients come from dustfall, dew, fog, and rainwater and are rapidly taken up by the thallus. When dry, lichens enter a dormant stage and can endure long periods of drought and great extremes of temperature. When moistened they are ready to resume growth at once. The result is that lichens have invaded and thrive in almost all habitats from deserts to the arctic tundra and the last exposed rock in Antarctica. They flourish on bare rock and other pioneer habitats where normal leafy green plants cannot gain a foothold. Often their only competitors are mosses. Because they remain dormant so much of the time and grow so slowly, lichens are the true Methuselahs among living organisms. Some arctic species are estimated to be 5000 to 9000 years old—nearly twice the age of bristlecone pines in California.

History of California Lichens

The first lichen collector in California was probably the explorer Archibald Menzies, who visited the Monterey area in 1792. Most of his collections are now preserved in the British Museum. From 1863 to 1875, H. N. Bolander collected widely in California and sent his material to Edward Tuckerman, the father of American lichenology, who described many new species unique to the state.

The first comprehensive study of California lichens was published by A. W. C. T. Herre (1910), who had a lifelong interest in lichens although his primary work was in fishes. He collected extensively in the Santa Cruz peninsula. At almost the same time a German physician, Hermann Hasse, who lived in Santa Monica and worked in a nursing home, published a guide to the lichens of southern California (1913), mainly Los Angeles to San Diego. Both of these books, however, lack illustrations and are highly technical. Moreover, they are now very much out of date.

No comprehensive lichen flora of California has yet been published. Although we have Fink's *Lichen Flora of North America* (1935), which treats most of the California lichens known at that time, it too is now out of date. *How to Know the Lichens* (Hale, 1979) includes California lichens but not in great detail. There are as well numerous smaller articles published in scientific journals.

The first catalog of California lichens was published by Tucker and Jordan in 1979. It is a compilation of all the older reports, listing in total 999 species. This is nearly one-third of all lichen species known in North America! Almost half of the crustose forms, however, have not been re-collected since the early 1900s. It is apparent that urban development, agriculture, and air pollution have decimated many species, especially in the southern part of the state.

Economic Uses of Lichens

One of the most interesting uses of lichens is in dyeing. Long ago the Klamath Indians gathered specimens of the chartreuse Wolf Mosses (*Letharia columbiana* and *L. vulpina,* called "Swa-u-sam") and boiled them in water. Porcupine quills soaked in the extract turned a beautiful canary yellow and were woven into their baskets. The most brilliant dyes come from *Roccella* (see fig. 61), a rare coastal lichen in southern California. It yields a deep blue dye when soaked in ammonia, and it is this dye which is used in litmus paper, an acid-base indicator in chemistry laboratories. This lichen was once collected on a small scale in Baja California before coal-tar dyes were discovered.

Many other lichen species, when available in sufficient quantity, can be boiled to extract reddish brown, yellow, or russet dyes. The longer the wool is soaked, the deeper the color. For more details one should refer to Bolton's text on lichen dyeing (1960).

Another commercial use that continues to this day in Europe is the collection and extraction of oakmoss (chiefly *Evernia prunastri* and *Ramalina* species) for use in perfumery. The oily extract is used as a fixative in fine perfumes. During the Second World War, when supplies from Balkan countries were cut off, a small oakmoss industry started up in Oregon and California but was later abandoned.

Lichens can be eaten and there are no poisonous species, except for Wolf Moss. We have no records of the use of lichens as food by native California Indians, but farther north in British Columbia other tribes collected and processed *Bryoria* into small cakes. The rock tripes (*Umbilicaria*) are not only edible but are considered a delicacy in Japan. They can be washed and added to salads or deep fried.

Lichens are important sources of food for snails, slugs, mites, springtails, and other invertebrates. These animals chew off the upper cortex and algal layer, often stopping short of the medulla if it contains unpalatable acids. The damaged lichens will regenerate but usually die off in areas where air pollution is a serious problem.

The antibiotic properties of lichens have long been recognized in folk medicine. Old Man's Beard (*Usnea*) is still used in China and was probably as popular with native California Indians as it is today with Seminoles in Florida. The active component is apparently usnic acid, the basis for the yellowish color of many lichens. It is used in antibiotic preparations in Europe under such names as Usniplant and Usno.

Vulpinic acid, the yellow pigment in Wolf Moss, is a mild poison which was used in the last century in northern Europe in bait to poison wolves. The Latin species name *vulpina* originated from this use.

Lichens have damaging effects which are not always appreciated. They are implicated in rock disintegration through the combined effects of mechanical penetration by the rhizines along with mineral chelation by dissolved lichen acids and accumulated calcium oxalate. This process is rather slow in resistant rocks—no more than a few millimeters a century—but can be more rapid in sandstones. Their damaging effect comes to our attention when they deface monuments or rock sites covered with Indian petroglyphs. These lichens can be controlled by applying fungicidal sprays.

Lichens are extremely sensitive to air pollution since the thallus accumulates pollutants efficiently but cannot excrete them. They can therefore provide a useful biological measure of air pollution—either by noting their presence or absence in and around cities or by collecting the colonies and measuring the concentration of lead and other toxic metals in the thalli. Several large-scale studies have already been made in the Los Angeles basin to measure the extent of air pollution.

Anyone interested in rock gardens will want to explore the possibility of adding lichen-covered rocks to the garden. It is best to collect whole, fairly large rocks and place them in sunny spots with good drainage. If there is little air pollution, the lichens will live on for many years and need no special attention.

Lichen Habitats and Collecting Sites

In the descriptions of species we will give general habitats where lichens are found according to Ornduff's (1974) classifi-

Woodlands

FIG. 2. Ornduff's classification of the woodlands of California.

cation of California plant communities (fig. 2). These communities are described as follows.

Coastal Scrub: This open community of low shrubs (not shown in fig. 2) occurs near the coast. It has relatively few lichens today because of development but is the ideal habitat for *Dendrographa, Heterodermia, Niebla, Parmotrema, Roccella,* and *Teloschistes.*

Valley and Foothill Woodland: This is dry, open, often savanna-like forest dominated by the various California oaks: Valley Oak, Coast Live Oak, Interior Live Oak, and others. It occurs in the foothills of the Sierra Nevada as well as the Coast Ranges, going up to about 4000 ft elevation. This community is often fenced off and used for cattle grazing. The most common lichen genera are *Flavoparmelia, Flavopunctelia, Melanelia, Physcia, Physconia, Pseudocyphellaria,* and *Ramalina.*

Montane Forest: This forest type is dominated by conifers (pines, White Fir, Douglas Fir, Incense Cedar, *Sequoiadendron,* and others) with local mixtures of California Black Oak and other broadleaf trees. It is found between 4000 and 6500 ft elevation and is especially common in the northern half of the state. Grazing is minimal, but logging and timber operations are being done on a large scale since this forest type lies primarily within National Forest lands. Characteristic lichens are *Bryoria, Cladonia, Hypogymnia, Letharia, Parmeliopsis,* and *Tuckermannopsis.*

North Coastal Forest: This community comprises the great redwood forests from the Santa Cruz Mountains along the North Coast Ranges to Del Norte County. It is a moist area with considerable fog and high rainfall. While the magnificent redwoods and Incense Cedar are the most conspicuous trees, there are extensive pockets of Bigleaf Maple, Oregon Ash, Tanbark Oak, Giant Chinquapin, Madrone, White Alder, and other broadleaf trees in and away from the fog zone. The major lichens are *Hypogymnia, Lobaria, Nephroma, Parmotrema, Sphaerophorus,* and *Usnea.*

Subalpine Forest: This community lies above Montane Forest at about 6500 ft elevation to the tree line, which varies from 9000 to 11,000 ft. The main trees are Limber and Lodgepole Pines and other pines. There is a deep snowpack at these elevations, and lichens on lower trunks and on soil are often smothered out. Lichens in general are poorly developed, the most conspicuous being *Ahtiana, Letharia,* and *Melanelia.*

Alpine Fell-Field: This zone (not shown in fig. 2) lies above the tree line in the Sierra Nevada, Trinity Mountains, and other high mountain areas. Here the lichens grow on exposed rocks, soil, or humus: *Peltigera,* saxicolous Physcias, *Pseudephebe, Rhizoplaca,* and *Umbilicarias.* There are no true arctic-alpine species such as one finds in Montana or Colorado.

Other communities listed by Ornduff do not represent unique habitats for lichens and are combined in this guide as follows: Chaparral is combined with Valley and Foothill Woodland; Montane Chaparral and Pinyon-Juniper are combined with Montane Forest. Few lichens actually survive in chaparral where burning is common.

Not many lichens are completely restricted to any one of these plant communities, and as anyone familiar with California plant life knows, sharp contrasts and unexpected, isolated communities can be found within a few miles of hilly terrain.

Richness of the lichen flora is best correlated with rainfall or fog, which is generally higher on west-facing slopes. The east faces of the Sierra Nevada, Coast Ranges, and White Mountains, which lie in rain shadows, are usually too dry for ideal lichen growth. The best areas for collecting lichens, then, are the Cascades, Klamath Mountains, west-facing slopes of the Coast Ranges, and lower to mid elevations in the western foothills of the Sierra Nevada.

Origins of the Lichen Flora

The higher-plant flora of California contains about 5000 species. Fully 30% of these are endemics—that is, they do not occur outside of the state. The lichen flora of about 1000 species has a far lower degree of endemism, although our knowledge of lichen distribution in California and the western states is still so poor in comparison with vascular plants that it is foolhardy to give actual figures. Among the approximately 300 species of foliose and fruticose lichens in California it appears that only *Bryoria spiralifera* and *Edrudia* are truly endemic, a very low 1% of the flora. On the other hand, many more crustose lichens have been described from the state and the degree of endemism could be much higher. For example, a recent treatment of the common crustose genus *Pertusaria* by Dibben

(1980) lists 14 species for California, of which 4 are endemic. Our knowledge of the taxonomy and distribution of other crustose lichens, however, is far too imperfect to be used as a base for similar comparisons.

The lichen flora of California consists of various floristic elements that have originated both within and outside of the state. The most important element occurs in the northern half of California and consists of many showy foliose and fruticose lichens in the genera *Alectoria, Bryoria, Cavernularia, Lobaria, Nephroma, Parmeliopsis, Placopsis, Platismatia, Pseudocyphellaria, Sticta, Tuckermannopsis,* and *Usnea.* In fact these lichens are almost always better developed in Oregon and Washington or the adjacent Rocky Mountains and are at their southern limit in California. In the southern part of the state— and also occurring abundantly in Baja California—are many genera typical of Mediterranean climates: *Dendrographa, Heterodermia, Niebla, Parmotrema, Roccella,* and *Teloschistes.* Finally, some important genera are better developed in California than in neighboring states: *Flavoparmelia, Flavopunctelia, Hyperphyscia, Hypogymnia, Parmelina, Physcia, Physconia,* and *Ramalina.*

Lichen Conservation

Lichens are not included in the Inventory of Rare and Endangered Plants of California. They are nevertheless subject to the same environmental stresses as vascular plants and if anything are more sensitive to air pollution than other plants. It is well established that two unusual fruticose lichens, *Teloschistes villosus* and *Trichoramalina crinita,* can no longer be found in California, although they still occur in Baja California. Recent studies in the San Gabriel and San Bernardino mountains have shown that less than half of the lichen species collected in the early 1900s by Hasse can be collected today. These lichens have apparently succumbed to air pollution or their special habitats have disappeared with development and urbanization.

It is fortunate that certain unusual lichen habitats are effectively protected by their inclusion in parks and reserves. The most outstanding examples are Point Lobos, Point Reyes, Patrick's Point, and the Channel Islands, where one can observe

numerous crustose species, such as orange Caloplacas, as well
as showy fruticose species such as *Dendrographa*, *Niebla*,
Roccella, *Schizopelte*, and *Teloschistes*.

A scientific inventory of rare and endangered lichens cannot
be made at this time since we have so few data on their distribution in the state. However, the following species certainly
fall into this category: *Cladina portentosa*, *Dendrographa leucophaea*, *Edrudia*, *Heterodermia erinacea*, *Hypogymnia mollis*, *Parmotrema austrosinense*, *P. hypoleucinum*, *Ramalina
duriae*, *Roccella* species, *Teloschistes exilis*, and *T. flavicans*.

How to Collect Lichens

Lichens are perennial plants and can be collected any time of
the year. Corticolous (bark-inhabiting) species can usually be
removed with a sturdy hunting knife. Saxicolous (rock-inhabiting) species, especially the crustose ones, must often be removed along with the rock substrate. Their removal requires
special care and experience with a stone chisel and geological
hammer. It is sometimes difficult to find good specimens because the rock fractures in the wrong direction; large chunks
must be broken down to manageable size. It is always important to collect enough material to allow for later losses incurred
when specimens are prepared for your collection. At the same
time part of the lichen should be left behind so that the colony
will regenerate.

After lichens are collected they can be put in paper bags if
dry. Never use plastic bags as the lichens will mold and rot
quickly if wet. Once dry, the specimens can be stored indefinitely. For final preparation, sort the specimens and place them
in folded packets of durable paper (fig. 3). Paste a 3 × 5 label
on the front of the packet and include data on locality, elevation, habitat notes, the collector, a number, and the date. As
your collection grows it can be stored vertically in shoe or file
boxes.

For the sake of greater neatness small specimens attached to
bark or rock can be glued on 3 × 5 file cards. Very bulky foliose and fruticose collections of *Cladonia*, *Letharia*, *Ramalina*,
Usnea, and the like should be wetted and pressed flat between
sheets of blotting paper and corrugated paper, as is done with
higher plants, and dried with a fan or several changes of blotters.

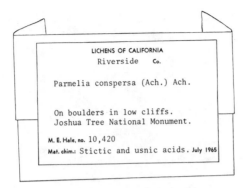

FIG. 3. Sample of a specimen packet made from $8\frac{1}{2} \times 11$ type-writer paper, with attached label showing typical data recorded.

After one collects a lichen specimen, the first question is: What is it called? Before that question can be answered, it is important to become familiar with the basic structure of lichens and the terms used to describe them. These terms, which are listed in the Glossary at the back of the book, are used in the identification keys and the descriptions again and again. They do not have to be learned all at once, but some terms must be thoroughly understood before attempting to use the keys. The outline of lichen structures given in the next section contains only the salient features, and students interested in further details are referred to more advanced texts (Hale, 1983; Lawrey, 1984).

Structure of Lichens

The basic lichen plant body is called a *thallus*. It consists mostly of fungal *hyphae*, threadlike fungal cells similar to those in common molds. In lichens the hyphae are organized into various kinds of tissues which can be distinguished under the microscope.

The typical orientation of tissues in a thallus is shown in fig. 4. The outermost tissue is the *upper cortex*, a dense layer of cells about 30 μm thick which protects the lichen. Under this cortex is a thin green *algal layer* consisting of colonies of green algae (usually *Trebouxia* or *Trentepohlia*). And below the algae is a loosely packed layer of hyphae, the *medulla*,

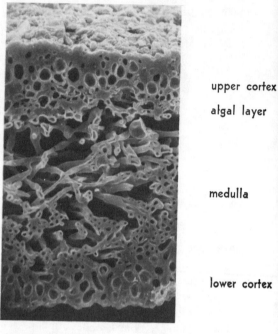

upper cortex

algal layer

medulla

lower cortex

FIG. 4. Cross section of a foliose lichen (*Phaeophyscia orbicularis*) showing tissue layers (×500).

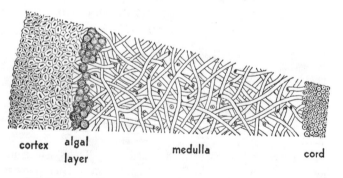

cortex   algal layer   medulla   cord

FIG. 5. Schematic cross section of a branch of *Usnea* sp. (magnified).

where foodstuffs are stored. Many foliose lichens also have a *lower cortex,* similar to the upper one. Fruticose lichens have a radial structure without a lower cortex (fig. 5).

There is one primitive group of foliose lichens which lacks any significant internal organization. The thallus consists of a single layer of hyphae intermingled with a blue-green alga, *Nostoc.* It is dark brown to bluish slate in color both externally and internally, and when moist it tends to swell up and appear gelatinous. In California this group includes *Collema* (pl. 3), *Hydrothyria,* and *Leptogium* (see fig. 28).

Growth Forms

While all lichens have a cortex, algal layer, and medulla, there are great differences in size and shape of the mature thallus. Lichens are traditionally divided into three growth forms: *foliose* (leaflike and growing more or less closely attached to bark, soil, or rock; see fig. 6); *fruticose* (hairlike or shrubby, dangling from tree branches, growing out as tufts from rocks or tree bark, or loose on soil; see fig. 7); and *crustose* (forming a crust on rocks, soil, or trees; see fig. 8). These growth forms may be recognized at sight without a hand lens. They do not

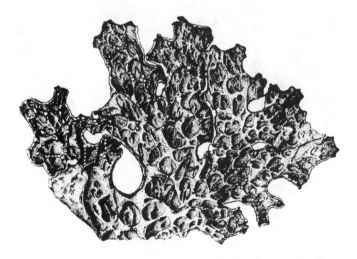

FIG. 6. Typical foliose lichen thallus (*Lobaria pulmonaria*) (×1).

FIG. 7. Typical fruticose lichen (*Ramalina menziesii*).

constitute a natural base for classifying lichens but are very convenient for purposes of lichen identification. In fact the California lichens in this guide are arranged according to growth form. Important features of each type are presented in appropriate sections.

FIG. 8. Typical crustose lichen thallus (*Caloplaca cerina*) (×10).

Vegetative Characters

Lichens produce a number of unique vegetative structures. These structures may occur in any one or more of the growth forms mentioned above. The most important ones are described here; unusual or exceptional features are discussed under the appropriate genera.

Soredia: These powdery grains consist of a small ball of hyphae about 50 to 100 μm in diameter enclosing a few algal cells. They function as vegetative propagules and are carried about by wind, water, or animals. They brush off on the fingers as a whitish powder. Soredia are produced in round or elongate masses called soralia, which occur on various parts of the thallus (fig. 9). Soralia lacking any orientation are called diffuse, as in *Cladonia* and *Lepraria*. It is essential to recognize soredia without hesitation, using a hand lens, since many lichen species are differentiated by their presence or absence.

Isidia: These are coralloid or fingerlike projections from the upper cortex, generally 0.3–1 mm high and about 0.2 mm in diameter, too small to be seen with the naked eye (fig. 9). They have no special orientation, occurring over the whole thallus

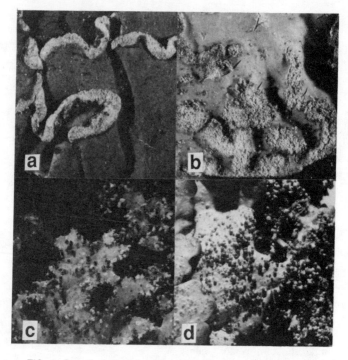

FIG. 9. Soredia and isidia: (*a*) marginal soredia of *Parmotrema austrosinense;* (*b*) laminal soredia of *Punctelia subrudecta;* (*c*) isidia of *Parmelia saxatilis;* (*d*) isidia of *Xanthoparmelia mexicana* (all ×10).

surface. They may also act as vegetative propagules when broken off. As with soredia, they are one of the most important characters used to identify species.

Cilia: These hairlike structures originate along the margins of lobes, especially in *Heterodermia* (fig. 10a), some Physcias, and *Parmelina.*

Rhizines: These are compact strands of hyphae 1–2 mm long produced on the lower cortex. They anchor the lichen to the substrate. They may be simple and unbranched or have various branching patterns (figs. 10 and 11).

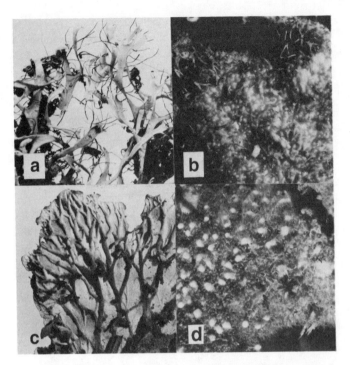

FIG. 10. Vegetative structures: (a) cilia of *Heterodermia leuco-melaena* (×3); (b) rhizines of *Punctelia subrudecta* (×10); (c) veins of *Peltigera membranacea* (×3); (d) tomentum (and raised papillae) of *Nephroma resupinatum* (×10).

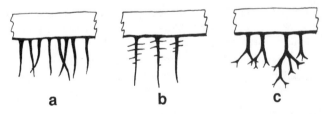

FIG. 11. Types of rhizine branching: (a) simple; (b) squarrose; (c) dichotomously branched.

Tomentum: This attachment organ differs from rhizines in consisting of separate, multicellular hyphae which resemble a layer of felt on the lower surface. It is characteristic of *Lobaria,* certain Leptogiums, *Nephroma, Pannaria, Pseudocyphellaria,* and *Sticta* (fig. 10).

Veins: These raised, veinlike structures occur on the lower surface of *Hydrothyria* and *Peltigera* (fig. 10). They are not related to the veins in plant leaves since they have no special function in translocation of nutrients or water.

Pores: Many lichens have small pores in the cortex for gas exchange. These pores are of two kinds: Cyphellae are relatively large, recessed pores (about 1 mm in diameter) occurring only in the lower cortex of *Sticta* (fig. 12a); pseudocyphellae are small round to angular holes (0.1–1.0 mm in diameter) in the upper or lower cortex (fig. 12b, c, d). Pseudocyphellae occur in *Alectoria, Bryoria, Melanelia, Parmelia, Pseudocyphellaria, Punctelia,* and *Sulcaria.* The white spots on some Physcias (fig. 12f) are not pores but merely algae-free spots below the cortex giving a mottled effect.

Pruina: This structure appears as a thin white frosty layer on the surface of many lichens, especially *Physcia* and *Physconia* species. It consists of accretions of calcium oxalate crystals (fig. 12e).

Cephalodia: These are small warty structures about 1 or 2 mm wide which consist of blue-green algae enclosed in cortical hyphal tissue (fig. 12g). They grow on *Peltigera, Placopsis,* and *Stereocaulon* and internally in some *Lobaria* species. Cephalodia have special interest because they can fix atmospheric nitrogen and thereby enrich the forest ecosystem.

Fruiting Bodies

The fungal component of most lichens forms fruiting bodies, spore-bearing structures involved in sexual reproduction. All of the lichens included in this guide are Ascomycetes (sac-producing fungi) and produce either apothecia or perithecia.

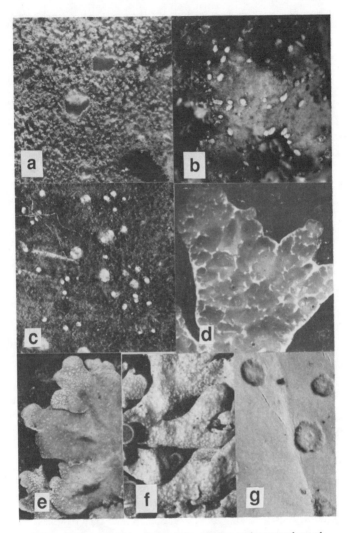

FIG. 12. Vegetative structures: (*a*) cyphellae on lower surface of *Sticta limbata;* (*b*) laminal pseudocyphellae of *Punctelia stictica;* (*c*) pseudocyphellae on the lower surface of *Pseudocyphellaria anthraspis;* (*d*) white markings (pseudocyphellae) on *Parmelia sulcata;* (*e*) pruina on the surface of *Physconia detersa;* (*f*) white spots on the surface of *Physcia aipolia;* (*g*) cephalodia of *Peltigera aphthosa* (all ×10).

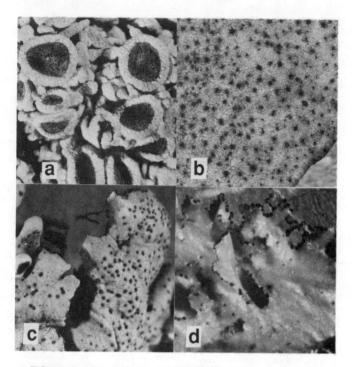

FIG. 13. Fruiting structures: (a) apothecia of *Physconia distorta;* (b) perithecia of *Dermatocarpon miniatum;* (c) laminal pycnidia of *Parmelina quercina;* (d) marginal pycnidia of *Tuckermannopsis pallidula* (all ×10).

Apothecia are generally small (1 – 10 mm wide) disk or cup-shaped structures on the thallus (fig. 13a). When sectioned vertically with a razor blade and examined under a microscope, the apothecia contain a fertile layer (the hymenium) of thread-like hyphae (paraphyses) and small sacs called asci (fig. 14). The ascus is the end product of the sexual reproductive process and usually contains eight microscopic spores. These spores are ejected from the fruiting bodies when wet and fall back on the rock or tree bark, forming a new lichen thallus if a suitable alga is encountered.

Perithecia contain the same tissues as apothecia but are immersed in the thallus as small flasks. Spores are ejected through

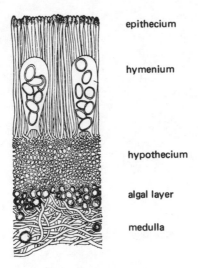

epithecium

hymenium

hypothecium

algal layer

medulla

FIG. 14. Cross section of a typical apothecium.

an apical pore at the surface of the thallus (fig. 13b). Except for squamulose *Catapyrenium* and foliose *Dermatocarpon*, only crustose lichens produce perithecia and are often called pyrenocarps.

Pycnidia are virtually identical in size and shape to perithecia and occur in all lichen groups. They contain not spores but hundreds of free bacilliform conidia, a kind of asexual spore which functions as male spermatia in some fungal groups (fig. 13c, d).

Apothecia and perithecia are very important in lichen taxonomy. They form a natural basis for family and generic classification. Crustose lichens cannot usually be identified without fruiting bodies.

Chemical Characters

Lichens contain a number of unique lichen acids, phenolic substances produced in no other plants although related to tannins in higher plants. The structures of these acids are now well known (fig. 15), and lichenologists routinely use them as aids to identify species since the components are usually constant.

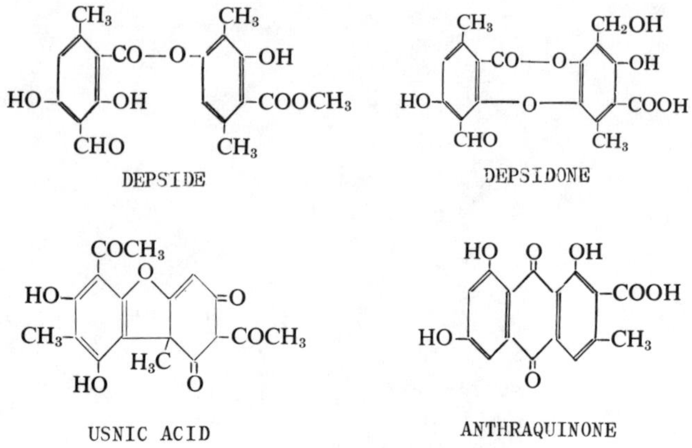

FIG. 15. Structural formulas of some lichen substances.

Very simple color reaction tests were developed in the 1870s and microcrystal tests were introduced in the 1930s as a means of making accurate identification of the acids. Now we use sophisticated technologies such as thin-layer chromatography (TLC) and high-pressure liquid chromatography (HPLC).

Fortunately, very few lichens need complete chemical analysis for identification. A combination of morphological and chemical characteristics revealed by the color tests is often adequate. A brief description of the techniques available is given here. Fuller details can be found in Hale (1983). Which of these techniques a beginning student will be able to use depends on the facilities available. Thallus color tests can be done at home and yield much useful information. Chromatography would have to be carried out in a laboratory equipped with exhaust hoods and necessary solvents and equipment.

Color Tests

Color tests were first discovered in the 1870s in connection with the preparation of lichen dyes. Certain lichen acids react yellow or red with potassium hydroxide (KOH, abbreviated K), others react red or rose with calcium hypochlorite (Ca(ClO)$_2$, abbreviated C), and a few react red with a combination of K followed by C. A third reagent, para-phenylenediamine (P), was added in 1936; it gives an instant yellow or red color reaction with certain acids.

Ordinary household bleach, such as Clorox, is an excellent substitute for calcium hypochlorite and can be used directly from the plastic bottle. For KOH we now use sodium hydroxide (NaOH), prepared by dissolving a stick of caustic soda in water (or purchased at drugstores in a 20% solution). Both K and C can be stored in stoppered bottles and kept indefinitely. (The C solution should be tested for strength if it has not been used for a week or more by applying a drop to a lichen species known to be positive, such as *Flavopunctelia flaventior*.) We do not recommend the use of para-phenylenediamine for beginners without access to a laboratory; it is a corrosive and potentially carcinogenic chemical.

Reagent K is applied to the cortex or to the medulla with a dropper or capillary tube. When a medullary test is called for, slice away a bit of the cortex with a razor blade to expose the medulla (fig. 16). The color change should be observed under a hand lens or binocular dissecting microscope. If positive it will be visible within a few seconds. The orange cortical pigment parietin in *Caloplaca, Teloschistes,* and *Xanthoria* reacts deep purple. Some of the other color reactions are as follows:

K+ yellow (+ indicates a positive reaction; − is negative): Atranorin (usually in the cortex only), alectorialic acid, stictic acid (both in the medulla).

K+ yellow turning red: Norstictic acid, salazinic acid (in the medulla only).

C+ rose or red: Gyrophoric acid, lecanoric acid, olivetoric acid (in the medulla only). Since this reaction may be rapid and fade quickly, it should be observed carefully.

P+ yellow, orange, or red: Fumarprotocetraric acid, norstictic acid, pannarin, protocetraric acid, psoromic acid, salazinic acid, stictic acid, thamnolic acid.

Microcrystal Tests

Microcrystal tests require a few simple reagents and a microscope. A fragment of the lichen about 1 cm across is extracted

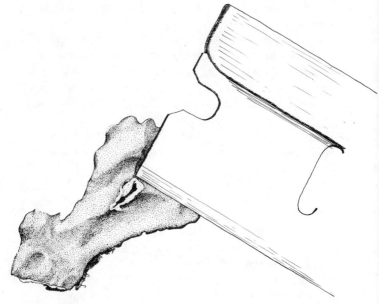

FIG. 16. How to section a thallus lobe to expose medulla for color testing.

with acetone on a microscope slide. After drying, a whitish ring will be seen and the thallus fragments are discarded. One or two drops of an appropriate crystallizing reagent (such as glycerine–acetic acid) are added and a cover slip is placed over this. The slide is warmed with a match or Bunsen burner until it barely begins to boil and is set aside to cool. The lichen acids crystallize out in unique shapes and colors. Unfortunately there are no complete discussions of this technique (see Hale, 1961 and 1979), which has value for beginners but has been superseded by chromatography.

Thin-Layer Chromatography

Thin-layer chromatography (TLC) is a standardized technique used since the 1960s for many plant and animal products. For analyzing lichens, a small piece of thallus is extracted with acetone in a small vial. The extract is spotted on a TLC plate (glass or aluminum-backed silica gel) and the plates are placed in tanks containing a solvent system (such as toluene–dioxane–acetic acid or hexane–ether–formic acid; see Hale, 1983, for more details). After the solvent front rises about 10 cm, the plate is removed from the tank, dried, sprayed with a 10% solution of sulfuric acid and heated at 100°C for about 5 minutes. The spots representing the acids will turn gray or yellowish. The color and position of these spots are used to identify the acids.

How to Use the Keys

Lichens, like other plants, can be identified in two ways: by comparing a specimen with a photograph or by running it through a key. Brilliantly colored orange and yellow lichens can often be identified with photographs quite accurately, but one must ultimately resort to keys to identify the more numerous greenish or whitish lichens which look superficially alike.

Keys are constructed of pairs of contrasting characters. Each numbered pair is read in turn until the specimen matches the description. The identification is confirmed or rejected by comparison with the photographs and the written description. A glossary of the technical terms is presented at the back of the book.

In this guide the lichens in California have been grouped into the three major growth forms: foliose, fruticose, and crustose. Under each group there is a discussion of the morphological features and a key to the genera. Then follows a detailed genus and species listing arranged in alphabetical order. Before long you will begin to recognize the main genera at sight and be able to bypass the genus key and go directly to the species listing.

Species enclosed in brackets in the keys are either species not given a separate entry and description (usually because they are rare) or species belonging to other genera which could also key out at this point and are discussed elsewhere in the guide. Data on elevation and habitats are based on our own field observations. Data on distribution of the species in California are compiled from our own and other collections but should not be considered exhaustive. There are still many new discoveries to be made in the state. County distribution maps, beginning on p. 239, are given for 40 of the more common species.

Key to Major Groups by Growth Form

1. Thallus foliose, leaflike with branching lobes, adnate to suberect, or umbilicate with a central umbilicus below . . . . Foliose Lichens (p. 27)
1. Thallus fruticose or shrubby, the branches round or flattened in cross section, tufted to pendulous and attached at the base or free-growing . . . . . Fruticose Lichens (p. 131)
1. Thallus crustose, closely attached to the substrate and lacking a lower cortex and rhizines but sometimes with a lobed margin; or thallus consisting of small crowded squamules . . . . Crustose Lichens (p. 176)

1 · FOLIOSE LICHENS

Foliose lichens occur on trees, rocks, and soil. They are the best-known group in California and are collected most frequently. They lie flat on the substrate and form broadly circular colonies which have distinct top and bottom sides. Individual colonies may vary in size from tiny appressed Physcias no more than 1 cm wide to huge, loosely attached Pseudocyphellarias 15–20 cm across.

The thallus is divided into branches called lobes, which are usually strap-shaped and 1–20 mm wide. Width of lobes is an important character and should be measured with a centimeter ruler, not guessed at (fig. 20).

Another basic character used for identifying specimens to genus and species is adnation, or attachment of the thallus. This feature ranges from very closely adnate to loosely attached. Adnate thalli cannot be removed easily from the substrate without breaking apart, while at the other extreme loosely attached specimens can be removed intact by hand or with a knife.

Color of the thallus is another critical character. Black, dark brown, and orange thallus colors do not pose any problems, but most lichens are light greenish to whitish mineral gray or some shade of greenish yellow. To distinguish between these colors, hold the lichen under a bright light or in the sun and compare it with a lichen of known color. For grays use any species of *Parmelia*, *Parmotrema*, or *Punctelia*. For yellows use the common *Flavoparmelia caperata*, *Flavopunctelia flaventior*, or any *Ramalina*, *Usnea*, or *Xanthoparmelia* species. As a general rule gray lichens react K+ yellow on the surface whereas yellow lichens show no reaction.

After you have determined thallus adnation, lobe width, and color, look for characters which require a hand lens: soredia, isidia, cilia, rhizines, tomentum, veins, and pores (see figs. 9 to 12). Various combinations of these characters delimit each genus so that you will soon be able to identify most lichen genera in the field.

Finally, some color tests with KOH and calcium hypochlorite may be called for, but these tests have to be done in the laboratory or at home. Microscopic examination of thallus structure, apothecia, and spores must also be done in the laboratory.

Key to Foliose Lichens

1. Thallus orange, greenish yellow, white, gray, or brown, the interior with a green algal layer and white medulla (cut open with a razor blade as in fig. 17 or scratch surface with fingernail).

    2. Thallus closely adnate to loosely attached and erect but clearly foliose and lobed, the lower surface rhizinate, tomentose, or bare. ....
    .... Group A: Stratified Foliose Lichens (p. 29)

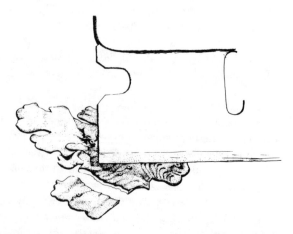

FIG. 17. How to section a lichen thallus to expose the inner layers.

   2. Thallus umbilicate, more or less round in outline and attached by an umbilicus below. ....
.... Group B: Umbilicate Lichens (p. 38)
1. Thallus dark or greenish brown to black or deep slate gray, the interior black, lacking any green or white layers. ....
.... Group C: Gelatinous Lichens (p. 38)

Group A: Stratified Foliose Lichens

1. Thallus orange or greenish yellow to yellow.
   2. Thallus orange, the surface reacting instantly K+ purple.
      3. Thallus chinky-crustose at the center with lobed margins; lower cortex and rhizines lacking. ......
.... *Caloplaca* (See Crustose Lichens, p. 176)
      3. Thallus foliose throughout with a distinct lower surface and rhizines. .... *Xanthoria*
   2. Thallus greenish yellow to yellow, K− or K+ very slowly yellowish.
      4. Center of thallus chinky-crustose, only the margins lobed (fig. 18); lower cortex and rhizines lacking; usually collected on rocks. ....
.... Crustose Lichens (p. 176)
      4. Thallus foliose throughout with a corticate white, brown, or black lower surface and sparse to moder-

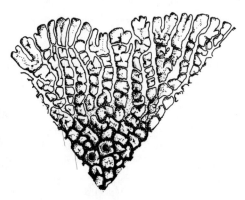

FIG. 18. Example of areolate, marginally lobate, crustose lichen growth form.

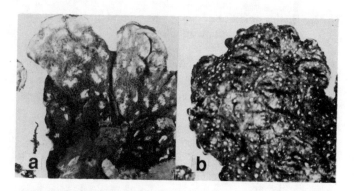

FIG. 19. Lower surface characteristics: (a) mottled tomentose (*Lobaria scrobiculata*); (b) pseudocyphellate tomentose (*Pseudocyphellaria anthraspis*) (×2).

ate rhizines (tomentose only in *Lobaria*); collected on trees, rocks, or soil.
5. Medulla lemon yellow (expose with razor blade). . . . . *Tuckermannopsis canadensis*
5. Medulla white throughout.
  6. Lobes 10–30 mm wide, the lower surface pale brown tomentose with large mottled spots (fig. 19a). . . . . *Lobaria*
  6. Lobes 0.5–10 mm wide, the lower surface white, brown, or black and rhizinate or bare.
    7. Upper surface with white pores (just visible without a hand lens); medulla C+ red. . . . *Flavopunctelia flaventior*
    7. Upper surface continuous, without pores; medulla C− (except in *Flavopunctelia soredica*).
      8. Lobes fairly broad, 2–10 mm wide, often loosely attached (see fig. 20).
        9. Margins of lobes with erect black pycnidia (see fig. 13d). . . . *Tuckermannopsis pallidula*
        9. Margins of lobes without pycnidia (or if present laminal and immersed in thallus).

 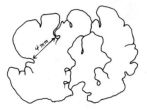

FIG. 20. How to measure lobe width.

10. Lower surface black. . . . .
Flavoparmelia caperata

10. Lower surface pale brown.

    11. Collected on rocks.
. . . . *Xanthoparmelia*

    11. Collected on trees.

        12. Apothecia present; medulla C−.
. . . . *Ahtiana sphaerosporella*

        12. Apothecia lacking; medulla C+ red. . . . .
Flavopunctelia soredica

8. Lobes narrow, 0.5–4 mm wide, closely adnate.

    13. Soredia lacking; almost always collected on rocks. . . . .
. . . . *Xanthoparmelia*

    13. Soredia present; collected on trees, rarely rocks.

        14. Collected on rocks. . . . .
. . . . *Xanthoparmelia mougeotii*

        14. Collected on trees.

            15. Lower surface white.
. . . . *Candelaria concolor*

           15. Lower surface black or brown.

                16. Soralia laminal and capitate; lower surface pale brown. . . . . *Parmeliopsis*

                16. Soralia marginal; lower surface black. . . . . *Hypotrachyna sinuosa*

1. Thallus white, greenish gray, brown, or blackish.

    17. Thallus branches inflated and hollow (cut open with a razor blade); lower surface black, lacking rhizines.

        18. Lower surface spongelike, with tiny pores (use hand lens) opening into large medullary cavities; lobes appearing flattened from above but inflated below. . . . . *Cavernularia*

        18. Lower surface entire or at most with large jagged holes, a black or white cavity running the length of the lobes.

            19. Apothecia usually present; soredia lacking. . . . . *Hypogymnia*

            19. Apothecia very rare; soredia always present.

                20. Upper surface with large, regularly spaced holes about 1 mm in diameter. . . . . *Menegazzia*

                20. Upper surface continuous or with only a few irregular holes. . . . . . . *Hypogymnia*

    17. Thallus branches flat and solid; lower surface white, brown, or black with tomentum or rhizines, rarely bare or lacking a cortex.

        21. Lower surface pale brown to whitish or mottled or with white pores or raised veins visible without a hand lens (figs. 10c, 19); thallus usually large, to 15 cm or more across, loosely attached and easily removed from the substrate.

        22. Lower surface with pale or dark veins (fig.

10c); collected on soil or base of trees. . . . .
. . . . *Peltigera*

22. Lower surface mottled or with white spots scattered in felty pale brown tomentum; collected on trees, rocks, or mosses over rocks.

    23. Lower surface mottled. . . . . . . *Lobaria*

    23. Lower surface with scattered white pores.

        24. Pores recessed, to 1 mm wide (use hand lens) (fig. 12a) . . . . *Sticta*

        24. Pores raised, less than 1 mm wide (fig. 19b).

            25. Apothecia present on lower surface of lobe tips; upper surface plane. . . . . *Nephroma*

            25. Apothecia (if present) laminal; upper surface reticulately ridged. . . . . *Pseudocyphellaria*

21. Lower surface lacking veins, pores, and tomentum, either brown, white, or black and rhizinate or bare (cortex lacking below in *Heterodermia*); thallus often small and closely adnate (except large and loose in *Nephroma, Parmotrema, Platismatia,* and *Tuckermannopsis*).

    26. Thallus white, greenish to mineral gray; upper surface distinctly K+ yellow (K− only in *Hyperphyscia, Phaeophyscia,* and *Physconia*).

        27. Upper surface with white angular markings (fig. 12d), barely visible without a hand lens. . . . . *Parmelia*

        27. Upper surface continuous, lacking pores or white markings (white-spotted under hand lens in some Physcias; fig. 12f).

            28. Lobes large, 3–20 mm wide, loosely attached to suberect and easily removed with a knife or hand (see fig. 20 for measuring).

                29. Lower surface rhizinate, at least toward the center, not mottled

    or wrinkled. . . . . *Parmotrema*
29. Lower surface bare, lacking rhizines or only very sparsely rhizinate.
  30. Lower surface shiny, becoming mottled black or brown and white. . . . . . .
    . . . . *Platismatia*
  30. Lower surface black, deeply and finely wrinkled. . . .
    . . . . *Esslingeriana*
28. Lobes small and narrow, 0.5–4 mm wide, adnate to closely appressed and usually not easily removed from the substrate without a knife (except for *Platismatia*).
  31. Lower surface black (lift up lobe with a razor blade or forceps and use hand lens).
    32. Lobes 2–3 mm wide; cortex K+ yellow.
      33. Lower surface rhizinate; apothecia usually numerous on lobe surface. . . . .
      Parmelina quercina
      33. Lower surface bare; apothecia if present marginal. . . . .
      . . . . . . *Esslingeriana*
    32. Lobes 1–2 mm wide; cortex K–.
      34. Upper surface pruinose and scabrid (fig. 12e). . . . . . *Physconia*
      34. Upper surface not pruinose. . . . .
      . . . . . . *Phaeophyscia*

31. Lower surface white or pale brown.
    35. Margins of lobes ciliate (fig. 10a) (hand lens not needed).
        36. Lower surface corticate and rhizinate, smooth and shiny. .... *Physcia*
        36. Lower surface lacking a cortex, dull and cottony or appearing fibrous. .... *Heterodermia*
    35. Margins of lobes smooth, lacking cilia.
        37. Thallus surface turning K+ yellow almost at once.
            38. Lower surface pale brown; usually collected on conifers. .... *Parmeliopsis*
            38. Lower surface white; usually collected on hardwoods or rocks, more rarely on conifers. .... *Physcia*
        37. Thallus surface K− or slowly dingy green.
            39. Thallus closely appressed; lower surface lacking rhizines (use hand lens). .... *Hyperphyscia*
            39. Thallus adnate; lower surface with numerous rhizines.
                40. Upper surface pruinose and scabrid; lobes 1–3 mm wide. .... *Physconia*
                40. Upper surface shiny, not pruinose; lobes 0.5–1 mm wide. .... *Phaeophyscia*

26. Thallus brown, ranging from light greenish to chestnut or olive blackish; upper surface K− or K+ slowly dull green.
    41. Lobes 3–20 mm wide, adnate to loosely attached.
        42. Apothecia present.
            43. Apothecia on lower surface of lobes.............*Nephroma*
            43. Apothecia along margins or on surface of lobes.
                44. Apothecia marginal; small erect pycnidia usually present (fig. 13d)..........
                    .... *Tuckermannopsis*
                44. Apothecia laminal; pycnidia (if present) laminal and immersed.........
                    .... *Melanelia*
        42. Apothecia lacking.
            45. Erect marginal pycnidia present (use hand lens)....
                .... *Tuckermannopsis*
            45. Pycnidia lacking (or if present laminal).
                46. Lower surface dark brown to black, rhizinate. ..... *Melanelia*
                46. Lower surface light brown, bare or very sparsely rhizinate.
                    47. Lower surface with sparse pale rhizines. ....
                        Tuckermannopsis
                    47. Lower surface completely bare.
                        ..... *Nephroma*
    41. Lobes 0.5–3 mm wide, narrow and crowded, closely adnate (suberect only in *Tuckermannopsis*).

48. Lobes suberect, with marginal pycnidia and apothecia. . . . . *Tuckermannopsis merrillii*
48. Lobes adnate to appressed; apothecia and pycnidia (if present) laminal.
    49. Lower surface bare or tomentose; lobes short and appearing almost squamulose; algae blue-green. . . . . *Pannaria*
    49. Lower surface rhizinate; lobes elongate, clearly foliose; algae green.
        50. Coarse pustules on lobe surface; always collected on rocks. . . . . *Neofuscelia*
        50. Pustules lacking but soredia or isidia may be present; collected on trees or rocks.
            51. Upper surface with white pores (visible without hand lens). . . . . . *Punctelia stictica*
            51. Upper surface without pores or with tiny pores visible only with hand lens.
                52. Thallus chestnut brown. . . . . *Melanelia*
                52. Thallus greenish brown. . . . . *Phaeophyscia*

Group B: Umbilicate Lichens

1. Thallus greenish yellow. .... *Rhizoplaca*
1. Thallus gray to dark brown.
    2. Thallus small, to 1 cm broad, with sorediate margins.
        .... *Peltula euploca*
    2. Thallus 2–8 cm broad, the margins entire to dissected but without soredia.
        3. Apothecia often present; thallus brown to blackish.
            .... *Umbilicaria*
        3. Perithecia present (see fig. 13b); thallus dull whitish gray. .... *Dermatocarpon*

Group C: Gelatinous Lichens

1. Collected submerged on rocks in small streams. ....
    .... *Hydrothyria*
1. Collected on trees, rocks, or soil, never submerged in water.
    2. Thallus dull, olive or blackish brown; surface plane or pustulate; lower surface similar to the upper in color.
        .... *Collema*
    2. Thallus usually shiny, slate-colored to blackish brown; lower surface similar to the upper or with white tufts of tomentum. .... *Leptogium*

Ahtiana Goward

This is a recent segregate from the genus *Parmelia,* characterized by a yellow thallus (usnic acid in the cortex), pale lower surface, lack of pseudocyphellae, and small spherical spores. There is only one species, which is almost always found on fir trees in exposed habitats at high elevation. It also occurs in the Cascades and northern Rocky Mountains into Canada.

Ahtiana sphaerosporella (Müll. Arg.) Goward   (pl. 1b)

Thallus yellowish green, closely adnate, 2–6 cm broad; lobes irregularly broadened, 2–5 mm wide, the surface finely wrinkled except along the margins, black pycnidia present; lower surface tan, moderately rhizinate with tan, simple rhizines. Apothecia common, disk 1–4 mm wide, greenish or light brown. Cortex and medulla K−, C−, P− (usnic acid, fatty acids).

Habitats: On conifers in North Coastal, Montane, and Subalpine Forest from 5000 to 8000 ft elevation.

Range: Widespread from Fresno County northward to Tehama, Del Norte, Lassen, and Siskiyou counties on the western slopes of the Sierra Nevada, Klamath, and Siskiyou mountains and Modoc Plateau.

Candelaria Müll. Arg.

This conspicuous yellow lichen has very small crowded lobes and could be mistaken for a crustose lichen. However, the lower surface, viewed with a hand lens, is corticate with short rhizines. Spores for the genus are one-celled and colorless. The single species in California is common and widespread on oaks and other broadleaf trees in the foothills of the Coast Ranges and the Sierra Nevada, mostly below 3500 ft elevation. It could be confused only with *Xanthoria candelaria,* a more orange-tinged lichen which reacts K+ purple. *Candelaria* is K−.

Candelaria concolor (Dicks.) B. Stein   (pl. 2a)

Thallus greenish yellow, closely adnate on bark, 0.5−1 cm broad but fusing into larger colonies and often appearing as a scurfy crust; lobes very narrow, 0.1−0.3 mm wide, margins becoming finely dissected and sorediate; lower surface white, sparsely rhizinate. Apothecia rather rare, 0.3−1 mm in diameter, disk yellow. Cortex and medulla K−, C−, P− (calycin).

Habitats: On Valley Oak, California Black Oak, and other broadleaf trees and conifers in North Coastal Forest, Valley and Foothill Woodland, and occasionally in Montane Forest from 500 to 6300 ft elevation.

Range: See fig. 72c.

Cavernularia Degel.

This curious lichen is similar to *Hypogymnia* in having inflated lobes but can be recognized by the finely pored lower surface. (Use a hand lens.) These pores are openings to large invaginated cortex-lined pockets in the medulla. The spores of the fertile species, *C. lophyrea,* are colorless and one-celled, as in *Hypogymnia.* A closely related sorediate species, *C. hultenii* Degel., has been collected in Del Norte, Humboldt, and Mendocino counties. Both species are found on conifers in the North Coastal Forest.

1. Thallus with conspicuous apothecia, lacking soredia. . . .
. . . . *C. lophyrea*
[1. Thallus lacking apothecia but with soredia. . . . .
. . . . *C. hultenii*]

Cavernularia lophyrea (Ach.) Degel.   (fig. 21a)

Thallus greenish gray, adnate, 1–4 cm broad; lobes to 1 mm wide, linear, somewhat inflated (especially when seen from be-

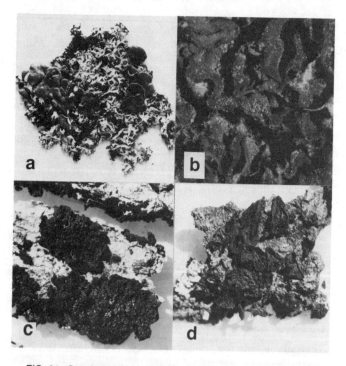

FIG. 21. Species of *Cavernularia* and *Collema:* (a) *Cavernularia lophyrea;* (b) *Collema crispum;* (c) *C. furfuraceum;* (d) *C. nigrescens* (all about ×1).

low) to more or less flattened, the surface shiny with black pycnidia; lower surface black, almost spongelike, with large perforations, dark within, rhizines lacking. Apothecia common,
2–7 mm in diameter, disk brown. Cortex K+ yellow; medulla
K−, C−, KC+ rose, P− (atranorin and physodic acid).

Habitats: On trunks of Douglas Fir, pine, and spruce in North
Coastal Forest near sea level.

Range: Rare from Mendocino County northward to Humboldt
County in the North Coast Ranges.

Collema Wigg.

Collema is a gelatinous (jellylike) lichen with a blue-green
phycobiont. The plant body is dull black and foliose, grading
into subcrustose, and lacks any internal organization into cortex and medulla. The apothecia, when present, have a dark
disk and colorless, septate or muriform spores which are important for species identification. The closest relative is *Leptogium,* similar in external appearance but having a shiny cellular cortex and often a bluish cast. When in doubt as to genus,
make a freehand cross section of the thallus with a razor blade
and examine it under the microscope for the presence or absence of a cortex.

There are about 12 species of *Collema* in California as recently reported in Degelius's (1974) monograph for North
America. A serious student must consult this work. *Collema* is
a difficult genus and it takes some experience to tell the species
apart. Since there are no known chemical substances, one must
rely on intergrading vegetative characters and the spores. We
can recognize three broad groups: corticolous species (two
treated here), saxicolous species (one treated), and soil-inhabiting species (one treated).

1. Thallus isidiate.
    2. Isidia flattened; thallus surface plane. . . . . . *C. crispum*
    2. Isidia cylindrical; thallus pustulate. . . . *C. furfuraceum*
1. Thallus smooth, lacking isidia.
    3. Collected on trees; thallus surface pustulate. . . . .
    . . . . *C. nigrescens*
    3. Collected on soil; thallus surface plane. . . . . . *C. tenax*

Collema crispum (Huds.) Wigg. (fig. 21b)

Thallus black, adnate, 2–3 cm broad; lobes 1–1.5 mm wide, somewhat ascending at the tips, the surface finely isidiate or lobulate-isidiate; lower surface smooth, gray, rhizines lacking. Apothecia common, 1–1.5 mm in diameter, disk reddish brown with a lobulate rim; spores four-celled, 15 × 30 μm.

Habitats: On rocks in sheltered areas in Valley and Foothill Woodland from near sea level to 2000 ft elevation.

Range: Los Angeles County northward to the San Francisco Bay area in the South Coast Ranges.

Collema furfuraceum (Arn.) DR. (fig. 21c, pl. 3b)

Thallus brownish black, adnate, 1–3 cm broad; lobes 2–4 mm wide, apically rotund, the surface broadly wrinkled and pustulate, isidiate, the isidia small, cylindrical; lower surface smooth, blackish, rhizines absent. Apothecia not seen.

Habitats: On oaks and other broadleaf trees and on rocks in North Coastal and Montane Forest and in Valley and Foothill Woodland from near sea level to 5000 ft elevation.

Range: See fig. 72d.

Collema nigrescens (Huds.) DC. (fig. 21d)

Thallus brownish green to black, adnate, 1–5 cm broad; lobes 3–5 mm wide, apically rotund, the surface ridged and pustulate; lower surface greenish brown, pitted, rhizines lacking. Apothecia very common, to 2 mm in diameter, disk brownish red with a narrow rim; spores colorless, transversely septate, 3–5 × 50–90 μm.

Habitats: On oaks and other broadleaf trees in North Coastal Forest and Valley and Foothill Woodland from 1000 to 4800 ft elevation.

Range: Rather common in San Diego County and from the Santa Cruz Mountains northward to Humboldt and Siskiyou

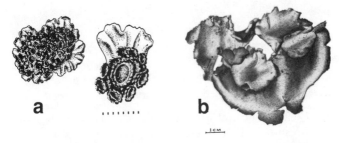

FIG. 22. Species of *Collema* and *Dermatocarpon*: (*a*) *Collema tenax* (scale in millimeters); (*b*) *Dermatocarpon miniatum* (×1).

counties in the North Coast Ranges and Klamath Mountains, rarer from Tulare County to Plumas County on the western slopes of the Sierra Nevada.

Collema tenax (Sw.) Ach. (fig. 22a)

Thallus closely attached to soil, blackish brown, consisting of short crowded lobes 1–2 mm long, the margins becoming covered with knobby warts, colonies 3–6 cm broad. Apothecia often present, 0.5–1 mm in diameter, the disk reddish brown with a dark rim; spores colorless, transversely septate with four or five locules, 15 × 24 μm, eight per ascus.

Habitats: On moist sandy banks or on soil over rocks in Valley and Foothill Woodland from sea level to 3000 ft elevation.

Range: Rare but probably overlooked from San Diego County northward to Mendocino County in the North and South Coast Ranges.

Dermatocarpon Eschw.

This genus consists of whitish or greenish brown, rock-inhabiting umbilicate lichens. They differ from *Umbilicaria*, with which they may be easily confused, in being pyrenocarpous—that is, the fruiting bodies are perithecia, not apothecia, immersed in the medulla and opening at the surface as black dots. Another group of species formerly classified in *Dermatocarpon* has a squamulose growth form and grows on soil; these are now placed in the genus *Catapyrenium* (see Crustose Lichens). Even with this more restricted delimitation, *Dermatocarpon* is not an easy genus because of intergradation in most characters used for identification, such as formation of papillae on the lower surface and convolution of lobes. Unfortunately it has never been monographed for the North American species.

[1. Collected on wet rocks partially submerged in lakes or streams; thallus crowded, turning green when wet. . . . . *D. weberi*]
1. Collected on boulders everywhere (but never submerged); thallus expanded, not turning green when wet.
   [2. Black, dotlike pycnidia and apothecia on upper surface; medulla always C+ rose. . . . . . . *Umbilicaria*]
   2. Perithecia forming black dots (see fig. 13b); apothecia lacking; medulla always C−.
      3. Lower surface dark, smooth. . . . . . *D. miniatum*
      [3. Lower surface dark to black, finely papillate. . . . . *D. reticulatum*]

Dermatocarpon miniatum (L.) Mann   (fig. 22b)

Thallus light brown to gray, loosely adnate, 1.5–5 cm broad, the surface plane to convolute and undulate, often with black pycnidia; lower surface brown to blackish, bare to finely papillate, lacking rhizines. Perithecia visible as small black dots on the surface. Cortex and medulla K−, C−, P− (no lichen substances present).

Habitats: On sheltered rock outcrops in Valley and Foothill Woodland, Montane Forest, and Subalpine Forest from 1000 to 8500 ft elevation.

Range: See fig. 73a.

A second common species, *D. reticulatum* Magn., has the same range and external appearance but the lower surface is coarsely reticulate-papillate as seen with a hand lens. A third, rarer species, *D. weberi* (Ach.) Mann, grows at or near the waterline in streams or ponds, has crowded, more nearly foliose lobes, and turns green when wet. It has been collected in Mariposa County.

Esslingeriana Hale & Lai

Esslingeriana is an endemic American genus with just one species, which occurs in Idaho, Montana, and Oregon, as well as in California. It has a jet black, deeply wrinkled lower surface without rhizines, characters which separate it from *Tuckermannopsis* (formerly *Cetraria*). As with other *Cetraria*-like genera, the apothecia and the erect pycnidia occur on the margins of the lobes. *Esslingeriana* grows most often on dead twigs of conifers at higher elevations.

Esslingeriana idahoensis (Essl.) Hale & Lai   (fig. 23)

Thallus greenish mineral gray, loosely adnate to suberect on bark, 5–9 cm broad; lobes rather elongate, 2–4 mm wide, surface wrinkled and cracked, margins with black pycnidia, becoming dissected; lower surface jet black, wrinkled, sparsely rhizinate. Apothecia common, 2–6 mm wide, disk brown. Cortex K+ yellow, C−, P+ faint yellow; medulla K−, C−, P− (atranorin and fatty acids; traces of endocrocin).

Habitats: On branches and trunk of pines and Douglas Fir in Montane Forest from 1000 to 5000 ft elevation.

FIG. 23. *Esslingeriana idahoensis* (×1).

Range: Rather rare from Lake County northward to Siskiyou County in the Klamath Mountains and from Calaveras County to Shasta County on the western slopes of the Sierra Nevada.

Flavoparmelia Hale

This genus was formerly included in *Parmelia* or *Pseudoparmelia*. It has rather broad, adnate lobes and lacks marginal cilia. There is only one species in California, the well-known *F. caperata,* a common yellow-green lichen in the Coast Ranges. The only very similar species is *Flavopunctelia flaventior,* which has white pores (pseudocyphellae) and reacts C+ red in the medulla.

Flavoparmelia caperata (L.) Hale   (pl. 4a)

Thallus light yellowish green, closely adnate to adnate, 6–12 cm broad; lobes 5–9 mm wide, apically rounded, often crowded, the surface smooth to wrinkled with diffuse laminal soredia; lower surface black with a narrow bare marginal brown zone, rhizines sparse. Apothecia lacking. Cortex K−; medulla K−, C−, P+ red (usnic, caperatic, and protocetraric acids).

Habitats: On various oaks, manzanitas, Douglas Fir, and cultivated trees, less commonly on rocks, in North Coastal Forest and Valley and Foothill Woodland from sea level to 3000 ft elevation.

Range: Common from San Diego County to Humboldt County in the North and South Coast Ranges but rarer from Madera County to Sacramento County on the western slopes of the Sierra Nevada.

Flavopunctelia (Krog) Hale

This is a conspicuous and widespread lichen genus in the lower-elevation oak forests throughout the state. It can be recognized by the large greenish yellow thallus and white spots (pseudocyphellae) on the surface (although absent in *F. soredica*). The two species known in California were formerly classified in *Parmelia* or *Punctelia*.

Flavopunctelia flaventior (Stirt.) Hale   (pl. 4b)

Thallus greenish yellow, adnate to loosely adnate, 5–9 cm broad; lobes 4–8 mm wide, apically rotund, surface with distinct pseudocyphellae, sorediate laminally and marginally; lower surface black with a broad naked brown zone at the margins, smooth to wrinkled, sparsely rhizinate. Apothecia very rare, 2–4 mm wide. Cortex K−; medulla K−, C+, KC+ red, P− (usnic and lecanoric acids).

Habitats: On Valley Oak, Interior Live Oak, and other broadleaf trees, rarely on rocks, in Valley and Foothill Woodland from near sea level to 3000 ft elevation.

Range: See fig. 74a.

Closely related *F. soredica* (Nyl.) Hale, rarely collected in the Coast Ranges, is only about half as large, lacks pseudocyphellae, and has marginal, crescent-shaped soralia.

Heterodermia Trev.

This narrow-lobed foliose genus was long classified with *Anaptychia*. As newly defined (atranorin lacking, spores thin-walled), however, *Anaptychia* does not occur in California. The two species of *Heterodermia* here lack a lower cortex and have long marginal cilia and subfruticose, suberect lobes. The spores are brown, thick-walled, and two-celled. They can be told from *Physcia* species such as *P. adscendens* and *P. tenella* by lack of a lower cortex and rhizines. The genus is quite localized in the state, and *H. erinacea* in particular can probably be considered endangered.

1. Lobes long and ascending with soredia on the lower surface. . . . . *H. leucomelaena*
1. Lobes shorter and lacking soredia but often with apothecia. . . . . *H. erinacea*

Heterodermia erinacea (Ach.) Hale   (fig. 24a)

Thallus whitish gray, loosely adnate, 1–5 cm broad; lobes 1–1.5 mm wide, linear, the margins long ciliate; lower surface white, cortex and rhizines lacking. Apothecia seen only in his-

FIG. 24. Species of *Heterodermia*: (a) *H. erinacea*; (b) *H. leucomelaena* (×1.5).

torical collections, to 1.5 mm in diameter, the disk black with a white rim. Cortex and medulla K+, P+ yellow, C− (atranorin and zeorin).

Habitats: On shrubs or rocks in Coastal Scrub near the seacoast.

Range: Rather rare from Santa Barbara County to the San Francisco Bay area (historical records from San Diego County).

Heterodermia leucomelaena (L.) Poelt   (fig. 24b)

Thallus whitish gray, loosely adnate, 3–9 cm broad; lobes 0.5–1.5 mm wide, linear, sometimes ascending at the tips, margins with long, squarrose cilia, sorediate with diffuse soredia scattered over the lower surface; lower surface white, cortex and rhizines lacking. Apothecia not seen. Cortex K+ yellow; medulla K+ yellow turning red, P+ pale orange (atranorin, salazinic acid, and zeorin). (Some specimens may react K− and contain only zeorin in the medulla.)

Habitats: On oaks and other trees, rarely rocks, in North Coastal Forest, Valley and Foothill Woodland, and Coastal Scrub from near sea level to 1500 ft elevation.

Range: Rather rare in San Diego County northward to Humboldt County in the North and South Coast Ranges.

Hydrothyria Russ.

This unique genus is the only foliose lichen which grows submerged in water. It does not look like a normal lichen. At first glance in the field it seems to be a small tufted alga, such as one sees along the seashore in California but here strangely out of range. When dried it resembles a *Leptogium*. It has been collected just a few times in the Sierra Nevada.

Hydrothyria venosa Russ. (fig. 25a)

Thallus bluish black, tufted and suberect, 2–8 cm broad, leathery when dry; lobes irregularly widened, 2–5 mm wide;

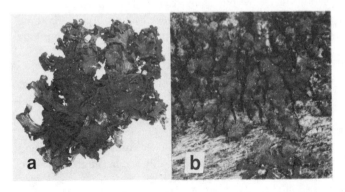

FIG. 25. Species of *Hydrothyria* and *Hyperphyscia:* (a) *Hydrothyria venosa* (×1); (b) *Hyperphyscia adglutinata* (×10).

lower surface dark with tannish veins, rhizines absent. Apothecia common, to 3.5 mm wide, disk brown. Cortex and medulla K−, C−, P− (no substances present).

Habitats: On stones or gravel submerged or partially submerged in small streams in Montane and Subalpine Forest from 5000 to 7000 ft elevation.

Range: Rare from Madera County to Calaveras County on the western slopes of the Sierra Nevada.

Hyperphyscia Müll. Arg.

Hyperphyscia (called *Physcia* or *Physciopsis* in older lists) was formerly classified in *Physcia* but differs in lacking rhizines and a lower cortex, in having long curved conidia (rod-shaped in *Physcia*), and in being very closely appressed to bark. It is closer to *Phaeophyscia,* especially *P. orbicularis,* which also lacks atranorin in the cortex (K−) but has rhizines below. All of these genera have brown, two-celled spores. The only species of *Hyperphyscia* in California, *H. adglutinata,* is quite inconspicuous, almost crustose in appearance, and is therefore probably overlooked.

Hyperphyscia adglutinata (Flk.) Mayrh. & Poelt (fig. 25b)

Thallus greenish to brownish gray, very closely adnate, less than 1 cm broad but fusing into larger colonies; lobes 0.3–0.5 mm wide, the tips shiny, soredia laminal, greenish, scattered; lower surface white, rhizines lacking. Apothecia sometimes developing, less than 1 mm in diameter. Cortex and medulla K−, C−, P− (no substances present).

Habitats: On Valley Oak, Interior Live Oak, and other broad-leaf trees in North Coastal Forest and Valley and Foothill Woodland from 1000 to 2000 ft elevation.

Range: Rare but probably overlooked from San Diego County northward to San Luis Obispo County and in Humboldt County in the North and South Coast Ranges.

Hypogymnia (Nyl.) Nyl.

This is one of the most interesting, commonly collected, and characteristic lichen genera in California. It is a whitish to greenish gray foliose lichen with long, narrow, inflated and hollow lobes. The lower surface is black and bare; the generic name translated literally from the Greek means "bare below." Apothecia are common and there are eight one-celled colorless spores. The only other similar genera are *Menegazzia,* with regular perforations on the upper surface, and *Cavernularia* with invaginated pores on the lower surface. Both of these genera are restricted to the North Coastal Forest. As a rule *Hypogymnia* grows on trunks and branches of conifers above the snowpack, more rarely on oaks and fenceposts at lower elevation.

For identification, it is first necessary to determine whether the interior of the lobes is white or dark by slicing them open with a razor blade. Soredia are readily identified, but the degree of adnation or extent of "trailing" lobes is sometimes difficult to determine. Much work, in fact, remains to be done before we achieve a complete understanding of *Hypogymnia* in California. At this time, a minimum of nine species is known. However, the common *H. imshaugii* group, species with a white interior, appears to include several undescribed populations.

1. Thallus sorediate; apothecia lacking.
    2. Soralia terminal, ring-shaped. . . . . *H. tubulosa*
    2. Soralia labriform or on the lobe surface.
        3. Soralia labriform at the lobe tips. . . . *H. physodes*
        [3. Soralia covering the lobe surfaces. . . . . *H. mollis*]
1. Thallus lacking soredia; apothecia usually present.
    4. Interior of lobes all white or roof only white (check several lobes).
        5. Interior all white . . . . *H. imshaugii*
        [5. Interior roof white, floor darkening. . . . .
            . . . . *H. metaphysodes*]
    4. Interior of lobes darkening to black.
        [6. Thallus surface regularly perforated with holes 0.5 mm in diameter; stictic acid present (K+ yellow). . . . . *Menegazzia terebrata*]

6. Thallus entire or with a few irregular perforations; stictic acid never present.

    [7. Lower surface with numerous fine perforations (use hand lens); collected only in North Coastal Forest. . . . . *Cavernularia lophyrea*]

    7. Lower surface smooth to rugose, at most with a few irregular perforations; collected in North Coastal Forest and elsewhere in the state.

        8. Lobes long and stringy, attached only basally.

            9. Branches linear and little branched, free, not appearing inflated; black lower surface conspicuous from above because of lobes twisted upward. . . *H. heterophylla*

            9. Branches shorter, usually strongly dichotomously branched and clearly inflated; black lower surface not conspicuous from above.

                10. Branches moderately inflated, 2–3 mm wide, uniformly thickened; medulla always P−. . . . *H. inactiva*

                10. Branches grossly inflated and irregularly thickened, 5 mm or more wide; P+ red or P−. . . . . *H. enteromorpha*

        8. Lobes elongate to short but generally adnate in places to bark throughout their whole length.

            [11. Roof of medullary cavity white. . . . . *H. metaphysodes*]

            11. Roof of cavity dark, same color as floor.

                12. Lobes loosely adnate, dispersed, often 5 mm or more wide, irregularly inflated with a black margin; medulla P+ red (more rarely P−). . . . . *H. enteromorpha*

                12. Lobes more adnate and crowded, usually less than 5 mm wide with-

out a black margin; medulla always P−...... *H. occidentalis*

Hypogymnia enteromorpha (Ach.) Nyl. (fig. 26a)

Thallus whitish gray to gray-green, adnate to loosely adnate, 5–10 cm broad; lobes 2–5 mm wide, strongly inflated with a black border, surface with black pycnidia; interior of lobes dark brown; lower surface black, shiny, wrinkled. Apothecia frequent, 1–14 mm wide, plane, and often cracked, disk brown. Cortex K+ yellow; medulla K−, C−, P+, or P− (atranorin, physodic acid, and physodalic acid with or without diffractaic and protocetraric acids; negative strain with only atranorin).

Habitats: On conifers, broadleaf trees, and fenceposts in the North Coastal Forest from sea level to 1600 ft elevation.

Range: See fig. 74b.

This was the first lichen to be described from California and was originally collected by Menzies in the late 1700s. In the past virtually all fertile Hypogymnias in California were called *H. enteromorpha*. This name is now limited to populations along the coast that are characterized by grossly inflated branches. It does not occur in the southern part of the state or in the Sierra Nevada, where it is usually misidentified as *H. imshaugii*.

Hypogymnia heterophylla Pike (fig. 26b)

Thallus whitish gray, loosely adnate to suberect, 3–7 cm broad; lobes 1–2 mm wide, linear and rather stringy, not strongly inflated, apically perforate, with a black border; interior of lobes brown to blackening; lower surface black, shiny, wrinkled. Apothecia common, 4–7 mm wide, disk brown. Cortex K+ yellow; medulla K−, C−, KC+ pink, P+ red (atranorin, physodic acid, and protocetraric acid).

Habitats: On pines and oaks in drier parts of the North Coastal Forest at lower elevations.

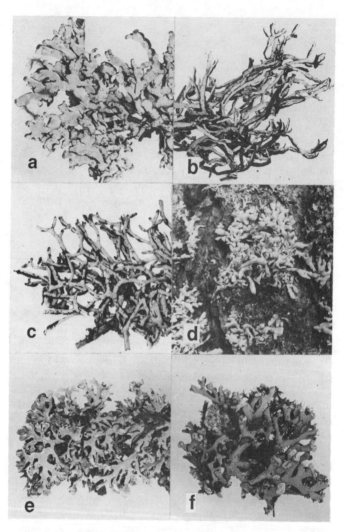

FIG. 26. Species of *Hypogymnia:* (a) *H. enteromorpha;* (b) *H. heterophylla;* (c) *H. inactiva;* (d) *H. occidentalis;* (e) *H. physodes;* (f) *H. tubulosa* (all about ×1).

Range: Rather rare from Marin County northward into Oregon in the North Coast Ranges.

Hypogymnia imshaugii Krog   (pl. 4c)

Thallus gray, loosely adnate to suberect, 5–10 cm broad, rather stiff; lobes 1.5–4 mm wide, rounded in cross section but not strongly inflated, the surface with numerous black pycnidia; interior of lobes snow white; lower surface black and wrinkled, becoming brown and smooth toward the tips. Apothecia very common, 2–13 mm wide, short-stalked, disk brown. Cortex K+ yellow; medulla K−, C−, P+ red, or P− (atranorin and physodic acid with or without diffractaic and protocetraric acids).

Habitats: On branches and trunks of oaks and other broadleaf trees and conifers in North Coastal Forest, Valley and Foothill Woodland, Montane Forest, and Subalpine Forest from 1000 to 8000 ft elevation.

Range: See fig. 74c.

This very common lichen varies considerably in lobe adnation. A high-elevation population contains only physodic acid (P− in the medulla) and is usually more closely adnate than specimens at lower elevation. Another population has adnate, broad but short, crowded lobes and is especially common in Sequoia National Park. *Hypogymnia metaphysodes,* collected rarely in Siskiyou and Trinity counties, differs in having a dark floor and white roof in the medullary cavity.

Hypogymnia inactiva (Krog) Ohlsson   (fig. 26c)

Thallus gray to gray-green, loosely adnate, 3–12 cm broad; lobes 1–3 mm wide, moderately inflated, tips perforate, surface with black pycnidia; interior of lobes brown; lower surface black to brown, weakly wrinkled. Apothecia frequent, 1–7 mm wide, disk brown. Cortex K+ yellow; medulla K−, C−, KC+ pink, P− (atranorin and physodic acid).

Habitats: On conifers in the North Coastal Forest from near sea level to 4500 ft elevation.

Range: Rather rare from the Santa Cruz Mountains northward into Oregon in the North Coast Ranges.

Hypogymnia occidentalis Pike (fig. 26d)

Thallus whitish gray, adnate to closely adnate, 3–12 cm broad; lobes 4–8 mm wide, short, crowded, surface with numerous black pycnidia; interior of lobes brown; lower surface black, shiny, deeply wrinkled. Apothecia common, 3–15 mm wide, disk brown, plane. Cortex K+ yellow; medulla K−, C−, KC+ pink, P− (atranorin and physodic acid).

Habitats: On conifers in the North Coastal and Montane Forest from near sea level to 5000 ft elevation.

Range: Rather rare from the San Francisco Bay area northward to Del Norte County in the North Coast Ranges and Klamath Mountains and in Plumas County on the western slopes of the Sierra Nevada.

Hypogymnia physodes (L.) Nyl. (fig. 26e)

Thallus light mineral gray, adnate to nearly suberect, 2–6 cm broad; lobes 0.5–2 mm wide, inflated, with black margins, sorediate in labriform patches on lower side of lobe tips that break open, black pycnidia present; medullary cavity variable, nearly white to brown; lower surface wrinkled, black but brown at the margins. Apothecia lacking. Cortex K+ yellow; medulla K−, C−, KC+ pink, P+ red (atranorin, physodic acid, and physodalic acid with or without protocetraric acid).

Habitats: On conifers and broadleaf trees or on mosses over rocks in the North Coastal Forest from near sea level to 2800 ft elevation.

Range: Rather rare from the Santa Cruz Mountains northward into Oregon in the North Coast Ranges.

Hypogymnia tubulosa (Schaer.) Hav. (fig. 26f)

Thallus light mineral gray, loosely adnate to suberect, 2–3 cm broad; lobes 1–1.5 mm wide, linear and little branched, moderately inflated, with terminal ring-shaped soralia; medullary cavity variably darkened; lower surface black and wrinkled. Apothecia lacking. Cortex K+ yellow; medulla K−, C−, KC+ pink, P− (atranorin and physodic acid).

Habitats: On twigs and branches of broadleaf trees and conifers or on mosses over rocks in North Coastal and Montane Forest and Valley and Foothill Woodland from near sea level to 3500 ft elevation.

Range: Widespread from San Benito County northward to Humboldt County in the North Coast Ranges and Klamath Mountains.

A third sorediate species in California, *H. mollis* Pike & Hale, is closest to *H. tubulosa* but is much smaller with diffuse soredia over the whole surface. It occurs rarely in San Luis Obispo County and San Diego County.

Hypotrachyna (Vain.) Hale

This primarily tropical genus is represented in California by two species, *H. revoluta* and *H. sinuosa,* both formerly classified in the genus *Parmelia. Hypotrachyna* is characterized by the dichotomously branched rhizines and narrow, eciliate lobes. Apothecial characters are the same as in *Parmelia.*

1. Thallus yellow-green (usnic acid present); medulla K+ red. .... *H. sinuosa*
[1. Thallus greenish gray (atranorin present); medulla K−, C+ red. .... *H. revoluta*]

Hypotrachyna sinuosa (Sm.) Hale (fig. 27)

Thallus pale yellowish green, adnate on bark, 2−7 cm broad; lobes 0.7−3 mm wide, sorediate, the soralia in rather diffuse marginal patches; lower surface black, moderately rhizinate, the rhizines dichotomously branched. Apothecia not seen. Cortex K−; medulla K+ yellow turning red, C−, P+ orange (usnic and salazinic acids).

Habitats: On branches and trunks of hardwood trees in North Coastal Forest from sea level to 1000 ft elevation.

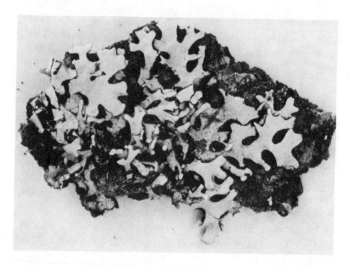

FIG. 27. *Hypotrachyna sinuosa* (×2).

Range: Humboldt County in the North Coastal Ranges.

This rare species is more common in Oregon and Washington. The second species, *H. revoluta* (Flk.) Hale, which contains gyrophoric acid, is also sorediate. It has been collected several times on trees and rocks from Santa Barbara County north to the San Francisco Bay area.

Leptogium (Ach.) S. Gray

Leptogium is a so-called gelatinous lichen with a blue-green phycobiont. While there is no internal organization and the medullary area is dark, there is a single-layered cellular upper and lower cortex (fig. 28). As a result, most of the species are shiny, in contrast to *Collema,* the other gelatinous lichen in the state, which lacks any cortical structure. However, it is not always easy to tell *Leptogium* from *Collema* without making careful vertical sections of the thallus with a razor blade and examining them under the microscope. Both genera have colorless, muriform spores.

There are at least 10 species reported from California, but many are represented by only a few historical collections. The

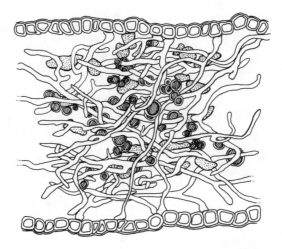

FIG. 28. Cross section of a *Leptogium* species (magnified).

species fall into two groups—those with a white tomentum below (*Mallotium*) and those with a bare lower surface hardly
distinguishable from the upper. Species identification with the
keys now available is not easy, especially since the morphological characters intergrade and there is no chemistry to fall
back on. The four species described here are rather common,
but it should be remembered that the treatment is incomplete.
Advanced students should refer to Sierk's (1964) monograph
on the North American species.

1. Thallus with white tomentum or tufts of haptera below.
    2. Lobes 2–5 mm wide, the margins with fine white hairs
       (use hand lens). . . . . *L. albociliatum*
    2. Lobes 4–10 mm wide, the margins without hairs.
       3. Surface finely wrinkled (use hand lens). . . . .
        . . . . *L. papillosum*
      [3. Surface smooth. . . . . *L. saturninum*]
1. Thallus bare below (or with a few tufts of haptera).
    4. Margins of lobes finely dissected and isidiate. . . . .
    . . . . *L. californicum*

FIG. 29. Species of *Leptogium*: (a) *L. albociliatum*; (b) *L. californicum*; (c) *L. palmatum*; (d) *L. papillosum* (about ×1).

4. Margins of lobes smooth to coarsely lobulate, lacking isidia.
   [5. Surface pustulate. . . . . *Collema nigrescens*]
   5. Surface smooth to wrinkled. . . . . . . *L. palmatum*

Leptogium albociliatum Desm.   (fig. 29a)

Thallus dark bluish gray to black, loosely adnate, 4–6 cm broad; lobes 3–5 mm wide, surface smooth with very fine white hairs toward the margins, usually isidiate, the isidia small and round; lower surface dark gray to black, smooth, with tufted white tomentum. Apothecia very common, less than 1 mm in diameter, disk reddish brown; spores one-septate, 8×15 μm.

Habitats: Among mosses on rocks or on mossy trail banks in Valley and Foothill Woodland and North Coastal and Montane Forest from 1000 to 6000 ft elevation.

Range: Rather common but easily overlooked from San Diego County northward to Humboldt County in the North and South Coast Ranges and from Mariposa County to Plumas County on the western slopes of the Sierra Nevada.

This distinctive species is sometimes placed in the genus *Leptochidium* Choisy.

Leptogium californicum Tuck.   (fig. 29b)

Thallus black to brownish black, adnate, 1–7 cm broad; lobes 1–2.5 mm wide, finely dissected with marginal isidia, surface finely wrinkled with a few isidia; lower surface dark gray, smooth, lacking tomentum. Apothecia rare, to 1 mm in diameter, disk reddish brown.

Habitats: On rocks and mosses over rocks in North Coastal Forest and Valley and Foothill Woodland from 1000 to 2000 ft elevation.

Range: Rather rare from Riverside County to Humboldt County in the North and South Coast Ranges and from Tulare County to Plumas County on the western slopes of the Sierra Nevada.

Leptogium palmatum (Huds.) Mont.   (fig. 29c)

Thallus brownish or chestnut gray, loosely adnate, 2–8 cm broad; lobes 1–3 mm wide, elongate and becoming convoluted at the tips, dissected with age, surface wrinkled; lower surface gray, wrinkled, tomentum lacking. Apothecia common, to 2 mm in diameter, disk reddish brown with a raised rim; spores muriform, about 15×40 μm.

Habitats: On rocks or mosses over rocks and soil in Valley and Foothill Woodland and North Coastal and Montane Forest from 500 to 4000 ft elevation.

Range: Fairly common from Riverside County northward to Humboldt County in the North and South Coast Ranges and from Fresno County to Plumas County on the western slopes of the Sierra Nevada.

Leptogium papillosum (B. de Lesd.) Dodge   (fig. 29d)

Thallus brownish black, loosely adnate, 3–6 cm broad; lobes 5–7 mm wide, quite rotund, the surface finely wrinkled with abundant isidia; lower surface light gray, covered with white tomentum. Apothecia very rare.

Habitats: On oaks and other broadleaf trees and rarely on rocks in North Coastal Forest and Valley and Foothill Woodland from 1000 to 3000 ft elevation.

Range: Rather common from Santa Barbara County to Humboldt County in the North and South Coast Ranges and from Tulare County to Plumas County on the western slopes of the Sierra Nevada.

This is the largest of the *Leptogium* species in California and typically grows on Valley, Garry, and Oregon Oaks. A very similar isidiate species, *L. saturninum* (Dicks.) Nyl., has a dull, smooth surface and occurs from the Santa Cruz Mountains to Humboldt County.

Lobaria Schreb.

Lobaria is one of the more showy lichen genera in northern California. The large foliose thalli have a mottled tomentose lower surface and a rugose to deeply foveolate surface. The algal symbiont is green or blue-green, depending on the species, and internal cephalodia with blue-green algae are present in some. The spores are colorless and one–nine septate.

There are four species in the state. One (*L. oregana*) is restricted to Humboldt and Del Norte counties; the others are more widespread from the Santa Cruz Mountains northward into Oregon.

1. Thallus distinctly yellowish (usnic acid present); lobe mar-

gins finely dissected and lobulate, soredia lacking. . . . . . .
. . . . *L. oregana*
1. Thallus greenish, dull yellowish, or pale brown; lobe margins lacking lobules but soredia present.
    2. Thallus surface strongly ridged and foveolate, greenish brown to brown, turning green when wet. . . . .
    . . . . *L. pulmonaria*
    2. Thallus weakly ridged, yellowish or ashy green, not turning deeper green when wet.
        3. Surface of lobes bare; medulla K+ yellow. . . . . . .
        . . . . *L. scrobiculata*
        [3. Surface with fine hairs (use hand lens); medulla K−. . . . . *L. hallii*]

Lobaria oregana Schreb. (fig. 30)

Thallus yellowish green, loosely attached to suberect, 8–12 cm broad; lobes 10–30 mm wide, deeply foveolate, the margins smooth but becoming dissected or lobulate at maturity; lower surface reticulately mottled, the bare areas yellowish tan, the tomentum dark brown. Apothecia not present. Cortex K−; medulla K+ yellow, C−, P+ orange (usnic, stictic, and norstictic acids).

FIG. 30. *Lobaria oregana* (×1).

Habitats: On trunk and canopy branches of conifers in moist woods in North Coastal Forest from sea level to 1000 ft elevation.

Range: Rare in Humboldt and Del Norte counties in the North Coast Ranges.

This lichen is much more common in neighboring Oregon and Washington, where it is called "Horse Lettuce," falling to the forest floor in great swatches.

Lobaria pulmonaria (L.) Hoffm. (fig. 6, pl. 5c)

Thallus brownish green, turning bright green when wet, loosely attached to suberect, 8–13 cm broad; lobes 6–10 mm wide, surface broadly ridged with some black pycnidia and coarse soredia along margins and ridges; lower surface light brown, tomentose with mottled white bare areas. Apothecia rarely found. Cortex K−; medulla K+ yellow, C−, P+ orange (stictic and norstictic acids).

Habitats: On oaks and other broadleaf trees and over mosses on trail banks in North Coastal and Montane Forest from sea level to 3000 ft elevation.

Range: See fig. 75b.

This conspicuous, unmistakable lichen has the common name Lungwort since it was used to treat tuberculosis in the Middle Ages.

Lobaria scrobiculata (Scop.) DC. (fig. 31)

Thallus yellowish green, loosely adnate, 4–15 cm broad; lobes little branched, 9–15 mm wide, the surface weakly ridged or foveolate with orbicular soralia along ridges; lower surface brown, tomentose with mottled bare white spots. Apothecia lacking. Cortex K−; medulla K+ yellow, KC+ reddish, P+ orange (usnic, norstictic, and stictic acids and scrobiculin).

Habitats: Base of trees and moss-covered rocks in North Coastal Forest from near sea level to 5000 ft elevation.

FIG. 31. *Lobaria scrobiculata* (×1).

Range: Rather rare from Santa Cruz County northward into Oregon in the North Coast Ranges.

An almost identical species, *L. hallii* (Tuck.) Zahlbr., has the same range as *L. scrobiculata*. It is distinguished by the fine translucent hairs on lobe tips.

Melanelia Essl.

The so-called brown Parmelias were recently separated from *Parmelia* because of the presence of brown pigments (Esslinger, 1977). The bulk of the species in North America have tiny pseudocyphellae and do not react when nitric acid is applied to the cortex. These species are referred to the genus *Melanelia*. A second brown genus, *Neofuscelia*, also represented in California, lacks pseudocyphellae, has more complex cortical structure, and reacts blue-green with nitric acid.

Melanelia has narrow to moderately broad, adnate lobes with black or brown rhizines below. It includes both corticolous and saxicolous species which are recognized chiefly by the presence or absence of soredia and isidia and by chemical reactions. All have colorless simple spores. Esslinger recognizes 12 species in California, and in fact *Melanelia* is one of the commonest foliose genera in the state. *Melanelia glabra,* in particular, is widespread and common on oak trees in Valley and Foothill Woodland.

1. Thallus sorediate.
  2. Soredia mostly marginal; lobes rather broad, 2–4 mm wide.
    3. Lower surface rhizinate; medulla C+ red. . . . . . .
    . . . . *M. subargentifera*
    [3. Lower surface bare; medulla C–. . . . .
    . . . . *Nephroma parile*]
  2. Soredia laminal; lobes narrow, 0.5–2 mm wide.
    4. Soredia mixed with fine isidia. . . . . *M. subaurifera*
    4. Soredia in pustules or powdery soralia without intermingled isidia.
      [5. Coarse soredia produced from large pustules.
      . . . . *Neofuscelia verruculifera*]
      5. Soredia powdery in capitate soralia, pustules lacking.
        6. Medulla C–. . . . . *M. disjuncta*
        6. Medulla C+ rose or red.
          7. Thallus uniformly brown; pseudocyphellae tiny, about 0.1 mm wide. . . . .
          . . . . *M. substygia*
          [7. Thallus greenish at lobe tips; pseudocyphellae large, to 0.5 mm wide. . . . . .
          . . . . *Punctelia stictica*]
1. Thallus lacking soredia.
  8. Thallus isidiate; apothecia rare.
    9. Isidia intermingled with soredia. . . . *M. subaurifera*
    9. Isidia not intermingled with soredia.
      [10. Medulla C+ red. . . . . *M. glabratula*]
      10. Medulla C–.
        [11. Isidia flattened and spatulate at maturity.
        . . . . *M. exasperatula*]
        11. Isidia mostly cylindrical to clavate.
          12. Isidia short, cylindrical; thallus usually becoming white pruinose. . . .
          . . . . *M. elegantula*
          [12. Isidia cylindrical to clavate but be-

                              coming dorsiventral and prostrate;
                              thallus rarely pruinose. . . . .
                              . . . . *M. subelegantula*]
    8. Thallus smooth to rugose, lacking isidia and soredia;
       apothecia common.
       13. Medulla C+ red.
           14. Thallus corticolous (rarely saxicolous); apo-
               thecia common. . . . . *M. glabra*
           [14. Thallus always saxicolous; apothecia usually
               lacking. . . . . *M. glabroides*]
       13. Medulla C−.
           15. Apothecia with 8 spores in ascus (examine
               sections under microscope). *M. subolivacea*
           [15. Apothecia with 16 spores in ascus. . . . .
               . . . . *M. multispora*]

Melanelia disjuncta (Erichs.) Essl.   (fig. 32a)

Thallus darkish brown, closely adnate, 2–4 cm wide; lobes narrow, 0.5–1.5 mm wide, the surface shiny, smooth to cracked,
sorediate, the soralia capitate, with coarse, darkening soredia;
lower surface black, moderately rhizinate. Apothecia not seen.
Cortex and medulla K−, C−, P− (perlatolic acid).

Habitats: On rocks in chaparral and coniferous forests in
Montane and Subalpine Forest from 1000 to 8000 ft elevation.

Range: Widespread in Riverside and San Bernardino counties, from the Santa Cruz Mountains to Trinity County in the
North Coast Ranges, and from Inyo County to Plumas County
in the Sierra Nevada.
    This blackish saxicolous lichen is often difficult to collect
since it grows on flat rock surfaces and being very closely adnate must be removed with the rock. A very similar species,
M. substygia (Räs.) Essl., has the same geographic range but
can be separated by a C+ rose medullary test (gyrophoric
acid). Closely related *Neofuscelia loxodes* and *N. verruculifera* could be confused with this species and all may be collected on the same rock. *Neofuscelia* has pustular outgrowths

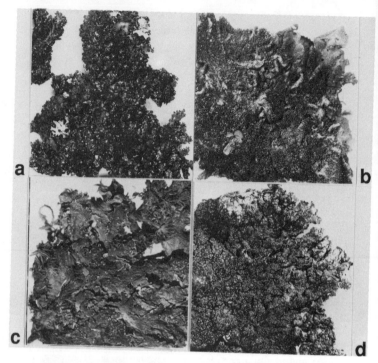

FIG. 32. Species of *Melanelia* and *Neofuscelia*: (a) *Melanelia disjuncta;* (b) *M. elegantula;* (c) *M. subargentifera;* (d) *Neofuscelia verruculifera* (×2).

without true soredia, although these capitate structures resemble soralia. If one has access to thin-layer chromatography, the *Neofuscelia* species are easily identified by the presence of divaricatic or perlatolic acids.

Melanelia elegantula (Zahlbr.) Essl.   (fig. 32b)

Thallus brown to olive green, loosely adnate to adnate, 2–10 cm broad; lobes 1–3 mm wide, surface becoming heavily white pruinose, isidiate, isidia laminal, cylindrical, branched with age, sometimes with apical white pseudocyphellae; lower surface black, turning brown toward the margin, moderately rhizinate. Apothecia not common, 2–7 mm in diameter, disk reddish brown with an isidiate rim. Cortex and medulla K−, C−, P− (no substances present).

Habitats: On oaks and occasionally rocks in Valley and Foothill Woodland and Montane Forest from 2000 to 8000 ft elevation.

Range: Widespread in Riverside County, from San Benito County to Siskiyou County in the North Coast Ranges and Klamath Mountains, and from Kern County to Modoc County on the western slopes of the Sierra Nevada and Modoc Plateau.

This is the commonest isidiate *Melanelia* in California. Two other isidiate species which also react C− are collected in northern California and may cause confusion: *M. exasperatula* (Nyl.) Essl., which has a shiny thin thallus and distinctly flattened, prostrate isidia, and *M. subelegantula* (Essl.) Essl., which has hollow dorsiventral isidia. Neither of these rare species develop white pruina to the extent of *M. elegantula.*

Melanelia glabra (Schaer.) Essl. (pl. 5d)

Thallus olive green to brown, closely adnate to adnate, 1−6 cm broad; lobes 2−3 mm wide, apically rotund, surface wrinkled toward the center, shiny at the margins; lower surface black, becoming brown at the margins, densely rhizinate. Apothecia very common, 2−5 mm in diameter, disk chesnut brown. Cortex K−; medulla K−, C+ red, P− (lecanoric acid).

Habitats: On Valley Oak, Interior Live Oak, California Black Oak, and conifers in Valley and Foothill Woodland and Montane Forest (rarely into Subalpine Forest) from 1000 to 6000 ft elevation.

Range: See fig. 75c.

This is one of the commonest lichens in the oak forests of California. The thallus is rather fragile and brittle when dry. The exposed medulla reacts instantly with C, and this color test is useful in separating it from the *M. subolivacea* group. A closely related C+ saxicolous species, *M. glabroides* (Essl.)

Essl., is collected in the Santa Cruz Mountains and Mariposa County.

Melanelia subargentifera (L.) Essl. (fig. 32c)

Thallus brown to olive brown, closely adnate, 4–8 cm broad; lobes 2–4 mm wide, apically rotund, sorediate, the soredia mostly marginal; lower surface black with brown margins, moderately rhizinate. Apothecia rare, 1–2 mm in diameter, disk brown. Cortex K−; medulla K−, C+ red, P− (lecanoric acid).

Habitats: On oaks and other broadleaf trees in Valley and Foothill Woodland and Montane Forest from 500 to 5000 ft elevation.

Range: Rare from Santa Barbara County northward to Tehama County in the North and South Coast Ranges and in Calaveras and Tuolumne counties on the western slopes of the Sierra Nevada.

The coarse, mostly marginal soredia resemble those of *Nephroma parile,* a rare species in Humboldt County with a bare, brown lower surface and C− medulla.

Melanelia subaurifera (Nyl.) Essl. (pl. 6a)

Thallus brown, closely adnate, 1–6 cm broad; lobes 2–3 mm wide, contiguous, sorediate on the surface, soralia diffuse and becoming finely isidiate, leaving a white cast when rubbed; lower surface light brown, moderately rhizinate. Apothecia rarely seen. Cortex K−; medulla K−, C+ red, P− (lecanoric acid).

Habitats: On oaks and other broadleaf trees and occasionally rocks in Valley and Foothill Woodland and North Coastal Montane Forest from near sea level to 6000 ft elevation.

Range: Rather rare from San Luis Obispo County northward to Humboldt County in the North and South Coast Ranges and Klamath Mountains and rare from Mariposa County to Shasta County in the Sierra Nevada.

A very close relative, *M. glabratula* (Lamy) Essl., has only fine isidia without soredia. It occurs in the North Coast Ranges north of San Benito and Contra Costa counties.

Melanelia subolivacea (Nyl.) Essl. (pl. 6b)

Thallus brown, closely adnate, 2–5 cm broad; lobes 1–2 mm wide, apically rotund, the surface wrinkled toward the center with shiny margins; lower surface brown, densely rhizinate. Apothecia very common, 1–3 mm wide, disk red-brown to brown; spores simple, colorless, eight per ascus. Cortex and medulla K−, C−, P− (no substances present).

Habitats: On oaks and other broadleaf trees and conifers in Valley and Foothill Woodland and North Coastal Montane Forest from near sea level to 6000 ft elevation.

Range: See fig. 75d.

This common, widespread *Melanelia* is distinguished by the almost constant presence of apothecia and the C− medullary test. The very similar *M. glabra* reacts C+ red. *Melanelia multispora* (Schneid.) Essl., a somewhat rarer species, is identical except for having 16 spores per ascus instead of 8. All fertile C− specimens should be examined microscopically. Finally, some specimens may become heavily papillate, nearly isidiate, and could be confused with species of *M. elegantula* group, which rarely produce apothecia.

Menegazzia Mass.

This rather rare lichen genus is represented by only one species in the Northern Hemisphere, *M. terebrata*. Externally it resembles *Hypogymnia* because of the inflated, hollow thallus lacking rhizines. *Menegazzia*, however, has small, fairly regular perforations over the upper surface, around which soredia may originate, a chemistry very different from *Hypogymnia*, and two to four large spores per ascus.

Menegazzia terebrata (Hoffm.) Mass. (fig. 33)

Thallus light gray, adnate, 2–4 cm broad; lobes 1–2 mm wide, swollen, dark at the margins, surface shiny, more or less regu-

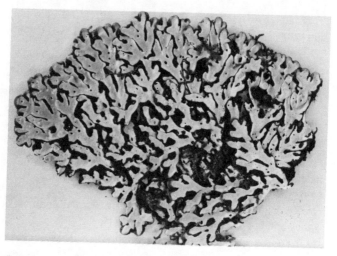

FIG. 33. *Menegazzia terebrata* (×1.5).

larly perforated with holes, sorediate, soralia powdery, origi-
nating around the holes or in erupting lobe tips; lower surface
black, wrinkled. Apothecia very rare. Cortex K+ yellow;
medulla K+ yellow and P+ orange (stictic acid and atranorin).

Habitats: On California Wax Myrtle, White Alder, and conifers in North Coastal Forest near sea level.

Range: Rare in Mendocino and Humboldt counties in the
North Coast Ranges.

Neofuscelia Essl.

This genus of small, brown, narrow-lobed saxicolous lichens
was formerly classified in *Parmelia*. It differs from other brown
Parmelias (*Melanelia*) in lacking pseudocyphellae and, in
California at least, in having pustular rather than powdery
soredia. The species often grow in rock crevices and are therefore difficult to collect. They have a complex chemistry, and
unfortunately it is difficult to separate the species accurately
without recourse to thin-layer chromatography. *Neofuscelia
verruculifera* is widespread in the state; *N. loxodes* (Nyl.)

Essl. and *N. subhosseana* (Essl.) Essl., both rarely collected, are confined to lava outcrops in the Modoc Plateau.

[1. Thallus with tiny pseudocyphellae; soredia powdery.
.... *Melanelia disjuncta*]
  1. Thallus lacking pseudocyphellae; coarse pustules present.
    2. Medulla KC+ rose (divaricatic acid); widespread in California. .... *N. verruculifera*
    2. Medulla KC− (divaricatic acid lacking); collected in the Modoc Plateau.
      [3. "Quintaria" unknowns on TLC. *N. subhosseana*]
      [3. Perlatolic acid on TLC. .... *N. loxodes*]

Neofuscelia verruculifera (Nyl.) Essl.   (fig. 32d)

Thallus dark brown, closely adnate, 1–4 cm broad; lobes to 1 mm wide, contiguous, surface shiny, finely wrinkled toward the center of the thallus, pustulate-isidiate, pustules coarse, clustered, bursting open and appearing sorediate; lower surface black, densely rhizinate. Apothecia lacking. Cortex K−; medulla K−, C−, KC+ rose, P− (divaricatic acid).

Habitats: On rocks in open sites in Valley and Foothill Woodland into Montane Forest from near sea level to 5000 ft elevation.

Range: Rather rare (but difficult to collect and probably overlooked) from San Diego County to Sonoma County in the North and South Coast Ranges and from Mariposa County to Siskiyou County on the western slopes of the Sierra Nevada and Modoc Plateau.

Nephroma Ach.

Nephroma is a brown, medium to broad-lobed lichen. It can be told from the brown Parmelias (*Melanelia* and *Neofuscelia*) by the brown, bare or finely tomentose lower surface and by apothecia, when present, found on the lower surface of lobe tips. The spores are colorless and two–four septate. Four species are known in California, one of which, *N. parile* (Ach.) Ach., is very rare and known only from Calaveras, Humboldt, and Siskiyou counties. It has marginal soredia and superficially re-

sembles *Melanelia subargentifera* (black below with rhizines).
All of the species occur at the base of trees or on mossy rocks
in the northern half of the state.

1. Lower surface with bare raised white spots scattered in the
   tomentum (hand lens not needed). . . . . . . *N. resupinatum*
1. Lower surface uniform, lacking spots.
   [2. Margins of lobes sorediate. . . . . *N. parile*]
   2. Margins smooth or isidiate-lobulate.
       3. Coarse lobules or isidia usually abundantly pro-
          duced; medulla white, K−. . . . . *N. helveticum*
       3. Lobules absent or only sparsely produced; medulla
          becoming pale yellow, reacting K+ pink or violet.
          . . . . *N. laevigatum*

Nephroma helveticum Ach. (fig. 34a)

Thallus light brown, loosely adnate, 3−4 cm broad; lobes 3−5
mm wide, apically rotund, the surface in part shiny, in part dull
tomentose, with some rounded isidia, margins with abundant
lobulate isidia; lower surface light brown, tomentose to nearly
bare. Apothecia on the lower surface of lobe tips, 4−7 mm
wide, disk brown. Cortex and medulla K−, C−, P− (unknown substances present).

a   b

FIG. 34. Species of *Nephroma:* (a) *N. helveticum;* (b) *N. laevi-*
gatum (×1).

Habitats: Base of trees or on mossy rocks in North Coastal and Montane Forest and Valley and Foothill Woodland from 1000 to 5000 ft elevation.

Range: See fig. 76a.

Nephroma laevigatum Ach. (fig. 34b)

Thallus light brown, loosely adnate, 4–8 cm broad; lobes rather broad, 3–5 mm wide, the margins entire; medulla pale yellow when exposed; lower surface tan, smooth. Apothecia common on lower surface of lobe tips, 2–8 mm wide, disk brown. Cortex K−; medulla K+ pink (nephromin and nephrin).

Habitats: Base of oaks and other broadleaf trees or on mossy rocks in North Coastal Forest from near sea level to 2500 ft elevation.

Range: Rather rare to locally abundant in the Santa Cruz Mountains northward to Humboldt County in the North Coast Ranges.

Nephroma resupinatum (L.) Ach. (pl. 6c)

Thallus brownish gray, loosely adnate, 3–7 cm broad; lobes 3–8 mm wide, apically rotund, surface becoming tomentose toward the tips; lower surface tan, densely tomentose with large scattered white spots. Apothecia common, on lower side of lobes, 3–9 mm wide, disk brown. Cortex and medulla K−, C−, P− (unidentified substances).

Habitats: On trees and shaded rocks in moist woods in North Coastal and Montane Forest from 400 to 4000 ft elevation.

Range: Widespread from the Santa Cruz Mountains northward to Del Norte County in the North Coast Ranges and in Calaveras and Plumas counties on the western slopes of the Sierra Nevada and in Shasta County in the Trinity Mountains.

Pannaria Del.

Pannaria is an inconspicuous, brown to bluish gray foliose to almost squamulose lichen with a blackish hypothallus around the thallus margin. The lower surface lacks a cortex and the phycobiont is blue-green. The apothecia have a thalline rim; spores are simple and colorless. The genus typically grows at the base of trees or on rocks in shaded habitats, but it is not collected often and probably overlooked. Six species have been reported from California, but some of the names will change when a revision of the genus is eventually completed. Only two species are discussed here.

Pannaria is typical of a widespread group of poorly known species classified in the family Pannariaceae. It includes in California the genera *Massalongia* Müll. Arg., *Pannaria, Parmeliella* Müll. Arg., and *Placynthium* (Ach.) S. Gray, all with a blue-green phycobiont and often collected on calcareous rocks and base of trees. They are keyed here without further discussion.

1. Apothecia lacking a distinct thalline rim.
    [2. Apothecia light brown. . . . . *Parmeliella*]
    [2. Apothecia black. . . . . *Placynthium*]
1. Apothecia with a well-developed thalline rim or apothecia lacking.
    [3. Margins of lobes with flattened isidia or lobules; always collected on mosses. . . . . *Massalongia carnosa*]
    3. Margins of lobes with cylindrical isidia; collected on trees, rocks, or rarely over mosses.
        4. Isidia becoming granular; apothecia rare. . . . .
            . . . . *Pannaria conoplea*
        4. Isidia distinct, not granular; apothecia usually present. . . . . *P. leucostictoides*

Pannaria conoplea (Ach.) Bory   (fig. 35a)

Thallus grayish brown to brown, closely adnate, 1–4 cm broad; lobes 1–1.5 mm wide, crowded, the margins coarsely isidiate to sorediate and sometimes white pruinose, the surface with a few isidia; lower surface black, tomentose. Apothecia rare, to

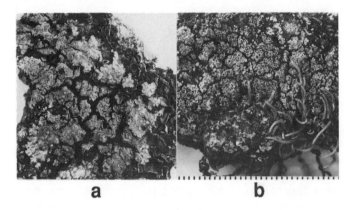

a **b**

FIG. 35. Species of *Pannaria:* (a) *P. conoplea;* (b) *P. leucostic-toides* (×2).

1 mm in diameter, disk brown to reddish, rim appearing sore-diate. Cortex K−; medulla K− and P+ orange-red (pannarin).

Habitats: Base of trees in Valley and Foothill Woodland from 1000 to 2000 ft elevation.

Range: Rather rare from the San Francisco Bay area north-ward to Sonoma and Trinity counties in the North Coast Ranges and in Plumas County on the western slopes of the Sierra Nevada.

Pannaria leucostictoides Ohlsson   (fig. 35b)

Thallus tannish gray, consisting of crowded squamules, 6−8 cm broad, edges of squamules ascending slightly, covered with coarse, white pruina-tipped isidia; lower surface dark, lacking a cortex. Apothecia common, 0.6−1.0 mm in diameter, disk dark brown with a white pruinose, crenulate rim. Cortex K−; medulla K−, C−, P− (zeorin).

Habitats: Base of trees in open forest in North Coastal Forest and Valley and Foothill Woodland from near sea level to 2000 ft elevation.

Range: Rare from Contra Costa and Sonoma counties northward to Tehama and Humboldt counties in the North Coast
Ranges and Klamath Mountains.

Parmelia Ach.

The large foliose genus *Parmelia,* as conceived by older workers like Fink (1935) and Herre (1910), is now considered to be
heterogeneous with many different elements. It has been divided into several smaller genera on the basis of cortical structure, type of rhizines and cilia, lobe width and adnation, and
conidial characters. As now delimited, *Parmelia* itself includes
only gray species with elongate pseudocyphellae in the upper
cortex. Other *Parmelia*-like genera in California include *Ahtiana, Flavoparmelia, Flavopunctelia, Hypotrachyna, Melanelia, Neofuscelia, Parmelina, Parmotrema, Punctelia,* and
Xanthoparmelia. A key to these genera follows.

1. Thallus yellow or greenish yellow.
    2. Pseudocyphellae on upper surface. . . . *Flavopunctelia*
    2. Surface continuous, without pores.
        3. Rhizines dichotomously branched. . . . .
        . . . . *Hypotrachyna sinuosa*
        3. Rhizines simple.
            4. Collected only on rocks.
                5. Lobes broad, 5–9 cm wide; coarse soredia
                present. . . . . *Flavoparmelia caperata*
                5. Lobes narrow, 1–3 mm wide. . . . .
                . . . . *Xanthoparmelia*
            4. Collected on trees, more rarely on rocks.
                6. Soredia lacking; collected on fir trees at high
                elevations. . . . . *Ahtiana*
                6. Soredia present; collected on oaks and other
                trees at lower elevations.
                    7. Soredia marginal. . . . .
                    . . . . *Flavopunctelia soredica*
                    7. Soredia laminal. . . . .
                    . . . . *Flavoparmelia caperata*
1. Thallus ashy to whitish gray or brown.
    8. Thallus dark brown. . . . . . *Melanelia* and *Neofuscelia*

8. Thallus ashy to whitish or greenish gray.
  9. Pseudocyphellae present on surface.
    10. Pseudocyphellae elongate. . . . . *Parmelia*
    10. Pseudocyphellae round. . . . . *Punctelia*
  9. Pseudocyphellae lacking.
    11. Lobes broad, 4–10 mm wide; thallus loosely attached. . . . . *Parmotrema*
    11. Lobes narrow, 2–3 mm wide; thallus adnate.
      12. Rhizines simple; lobe margins ciliate. . . .
      . . . . *Parmelina*
      12. Rhizines dichotomously branched. . . . . . .
      . . . . *Hypotrachyna*

The three *Parmelia* species treated here are generally narrow to medium-lobed and adnate. When fertile, all have simple, colorless spores. They occur on a variety of substrates and are more common in the northern half of the state.

[1. Pseudocyphellae round; lower surface pale or black; medulla C+ rose or red. *Punctelia*]
 1. Pseudocyphellae angular; lower surface always black; medulla always C−.
  2. Soredia more or less linear along ridges; medulla K+ red. . . . . *P. sulcata*
  2. Soredia coarsely isidiate or granular-isidiate on the surface; medulla K+ red or K−.
    [3. Rhizines richly branched (use hand lens). . . . . . .
    . . . . *P. squarrosa*]
    3. Rhizines simple to weakly branched.
      4. Medulla K+ turning red. . . . . . . *P. saxatilis*
      [4. Medulla K−, not turning color. . . . .
      . . . . *P. kerguelensis*]

Parmelia saxatilis (L.) Ach.   (fig. 36a)

Thallus whitish gray to gray, adnate to loosely adnate, 3–10 cm broad; lobes 2–6 mm wide, linear, surface somewhat ridged with angular white markings, coarsely granular-isidiate along the ridges and margins; lower surface black, densely rhizinate, rhizines black, simple. Apothecia very rare. Cortex

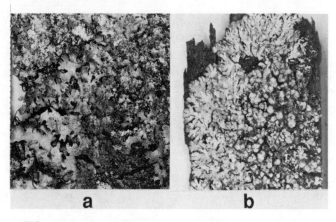

FIG. 36. Species of *Parmelia* and *Parmeliopsis:* (a) *Parmelia saxatilis;* (b) *Parmeliopsis hyperopta* (×2).

K+ yellow; medulla K+ yellow turning red, C−, P+ orange (atranorin and salazinic acid).

Habitats: On rocks, conifers, and broadleaf trees in North Coastal Forest, Montane Forest, and Valley and Foothill Woodland from near sea level to 6000 ft elevation.

Range: See fig. 76b.

An almost identical species rarely collected in the North Coastal Forest, *P. squarrosa,* has densely squarrosely branched rhizines as in *P. sulcata.* Another rare, closely related species is *P. kerguelensis* Cromb., which contains protocetraric acid and has no reaction with KOH.

Parmelia sulcata Tayl. (pl. 7a)

Thallus whitish gray to gray, adnate to loosely adnate, 4–8 cm broad; lobes 2–6 mm wide, linear, surface ridged with white markings; soredia powdery, marginal and along the ridges; lower surface black, rhizines dense, sparsely to densely squarrosely branched. Apothecia lacking or rarely present, adnate, 3–4 mm in diameter. Cortex K+ yellow; medulla K+ yellow turning red, C−, P+ orange (atranorin and salazinic acids).

Habitats: On broadleaf trees and conifers or on rocks in North Coastal Forest, Valley and Foothill Woodland, and Montane and Subalpine Forest from near sea level to 8000 ft elevation.

Range: See fig. 76c.

This is one of the commonest of the gray Parmelias in California. It can be told from closely related *P. saxatilis* by the powdery soredia and squarrosely branched rhizines. *Parmelia saxatilis* may have in part sorediate isidia, but the rhizines are unbranched.

Parmelina Hale

Parmelina is still another segregate of the old-form genus *Parmelia*. It is easily distinguished from *Parmelia* itself (for example, *P. saxatilis* and *P. sulcata*) by the lack of pseudocyphellae and the C+ red reaction. It differs from *Parmotrema* in the generally adnate, narrow lobes and from *Flavoparmelia* in the presence of marginal cilia and lack of usnic acid. There is only one species in California, *Parmelina quercina,* one of the most characteristic foliose lichens on California Black Oak.

Parmelina quercina (Willd.) Hale   (pl. 7b)

Thallus light mineral gray, rather closely adnate, 3–8 cm broad; lobes 2–3 mm wide, contiguous to crowded, axils with short black cilia on the margin; lower surface black, densely rhizinate, rhizines short, simple to sparsely branched. Apothecia common, laminal, 2–5 mm wide, disk brown. Cortex K+ yellow; medulla K−, C+, KC+ red, P− (atranorin and lecanoric acid).

Habitats: On oak trees in Valley and Foothill Woodland into Montane Forest from 500 to 5500 ft elevation.

Range: See fig. 76d.

Parmeliopsis (Stzb.) Nyl.

This narrow-lobed foliose genus is often confused with *Parmelia,* but it has different pycnidial formation. The two species in California are sorediate and externally identical except for the thallus color. Both grow typically at the base of Douglas Fir and other conifers, often below the snowpack. They are not likely to be confused with any other lichens.

1. Thallus whitish gray. *P. hyperopta*
1. Thallus yellow-green.
   2. Collected on trees. . . . . *P. ambigua*
   [2. Collected on rocks. . . . . . *Xanthoparmelia mougeotii*]

Parmeliopsis ambigua (Wulf.) Nyl.   (pl. 7c)

Thallus light yellowish green, closely adnate, 2–5 cm broad; lobes narrow, 0.6–0.8 mm wide, linear, surface smooth with capitate, laminal soralia; lower surface black with a narrow brown zone at the margins, moderately rhizinate. Apothecia lacking. Cortex and medulla K−, C−, P− (usnic and divaricatic acids).

Habitats: On conifers in North Coastal and Montane Forest from 2800 to 6000 ft elevation.

Range: Widespread from Sonoma County northward to Humboldt and Siskiyou counties in the North Coast Ranges and in the western foothills of the Sierra Nevada from Calaveras County to Shasta County.

   Xanthoparmelia mougeotii, recently discovered in Humboldt, Siskiyou, and Plumas counties, is superficially close to *P. ambigua* but is collected only on rocks and has a K+ yellow medulla (stictic acid).

Parmeliopsis hyperopta (Ach.) Arn.   (fig. 36b)

Thallus light gray, closely adnate, 2–4 cm broad; lobes narrow, 0.5–1 mm wide, surface with capitate, laminal soralia, lobe tips sometimes with pycnidia; lower surface brown, moderately rhizinate with short rhizines. Apothecia lacking. Cortex K+ yellow; medulla K−, C−, P− (atranorin and divaricatic acid).

Habitats: On conifers (especially Douglas Fir) in North Coastal and Montane Forest from 2000 to 5000 ft elevation.

Range: Widespread in Tehama, Trinity, Siskiyou, and Humboldt counties in the North Coast Ranges and Klamath Mountains and in Calaveras County on the western slopes of the Sierra Nevada.

Parmotrema Mass.

Parmotrema is a conspicuous foliose genus formerly included in *Parmelia*. It is characterized by broad, often suberect and ciliate lobes with a distinct bare marginal zone below. It lacks pseudocyphellae. The seven species in California, all confined to the Coast Ranges, are separated by presence or absence of isidia and cilia and by color reactions in the medulla.

1. Thallus surface isidiate. . . . . *P. crinitum*
1. Thallus sorediate.
    2. Cilia lacking; medulla C+ red. . . . . . *P. austrosinense*
    2. Cilia present on lobe margins; medulla C−.
        [3. Lower surface with a broad white zone at the margins. . . . . *P. hypoleucinum*]
       3. Lower surface dark brown at the margins.
          [4. Medulla K−. . . . . *P. arnoldii*]
         4. Medulla K+ yellow or red.
            5. Surface of thallus finely cracked or white reticulate (use hand lens and examine lobe tips). . . . . *P. reticulatum*
            5. Surface of thallus continuous or coarsely cracked only on older parts.
               6. Soralia mostly marginal and linear; lobes suberect. . . . . *P. stuppeum*
               6. Soralia broader, submarginal, the lobe tips turning down. . . . . *P. chinense*

Parmotrema austrosinense (**Zahlbr.**) **Hale**   (fig. 37a)

Thallus whitish gray, loosely adnate, 3–5 cm broad; lobes 5–10 mm wide, suberect, sorediate, the soralia marginal; lower surface black at the center but with a broad white margin 1–5 mm wide, sparsely rhizinate. Apothecia lacking. Cortex K+ yellow; medulla K−, C+ red, P− (atranorin and lecanoric acid).

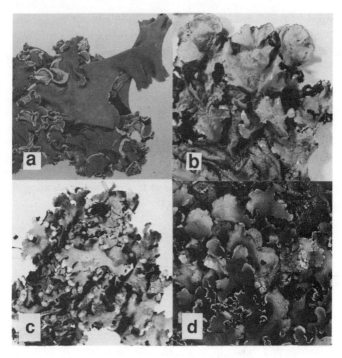

FIG. 37. Species of *Parmotrema:* (*a*) *P. austrosinense;* (*b*) *P. crinitum;* (*c*) *P. reticulatum;* (*d*) *P. stuppeum* (×1).

Habitats: On oak trees in Valley and Foothill Woodland from sea level to 2000 ft elevation.

Range: Rare in Santa Barbara County (historical range southward to Los Angeles County) in the South Coast Ranges.

Another rare but conspicuous species with a white lower surface, *P. hypoleucinum* (Stein.) Hale, has long cilia and reacts C− and K+ yellow (stictic acid). It occurs near the coast from San Diego to San Luis Obispo County.

Parmotrema chinense (Osbeck) **Hale & Ahti** (pl. 7d)

Thallus mineral gray, adnate to loosely adnate, 3–12 cm broad; lobes 3–6 mm wide; margins short ciliate, sorediate, soralia produced on revolute margins and lobe tips; lower surface black with a narrow bare brown zone, rhizines black, fairly dense. Apothecia absent, very rarely present, stalked, 5–8 mm in diameter. Cortex K+ yellow; medulla K+ yellow, C−, P+ orange (atranorin and stictic acid).

Habitats: On broadleaf trees and conifers, rarely on rock, in North Coastal Forest from near sea level to 2500 ft elevation.

Range: See fig. 77a.

This common species was known as *P. perlatum* until recently. A closely related, rather common coastal species, *P. arnoldii* (DR.) Hale, is virtually identical but reacts K− (alectoronic acid present) and has soredia concentrated more on short marginal lobes.

Parmotrema crinitum (Ach.) Choisy (fig. 37b)

Thallus whitish mineral gray, loosely adnate, 4–10 cm broad; lobes 3–5 mm wide, crowded, surface smooth with a few cracks, margins short ciliate, isidia laminal, abundant, becoming branched and ciliate at the tips; lower surface black with a narrow bare brown margin, rhizines fairly dense, black. Apothecia lacking. Cortex K+ yellow; medulla K+ yellow, C−, P+ orange (atranorin and stictic acid).

Habitats: On broadleaf trees and rocks in North Coastal Forest from near sea level to 1000 ft elevation.

Range: Rather rare in Humboldt and Del Norte counties in the North Coast Ranges.

Parmotrema reticulatum (Tayl.) Choisy (fig. 37c)

Thallus mineral gray (turning dull reddish when improperly dried), adnate to loosely adnate, 4–7 cm broad; lobes 3–6 mm wide, crowded, surface finely reticulate-maculate or cracked (under lens), with coarse submarginal soralia; margins sparsely ciliate; lower surface black, rhizines black, sparse. Apothecia lacking. Cortex K+ yellow; medulla K+ yellow turning red, C−, P+ orange (atranorin and salazinic acid).

Habitats: On oaks and other broadleaf trees in Coastal Scrub and Valley and Foothill Woodland at lower elevation.

Range: Widespread from Los Angeles County to Santa Barbara County and the Channel Islands in the South Coast Ranges.

Parmotrema stuppeum (Tayl.) Hale (fig. 37d)

Thallus mineral gray (turning dull reddish when poorly preserved), loosely attached, 5–9 cm broad; lobes 4–8 mm wide, little branched, margins long ciliate with narrow marginal soralia; lower surface black with a bare brown marginal zone, rhizines black, sparse. Apothecia rarely seen. Cortex K+ yellow; medulla K+ yellow turning red, C−, P+ orange (atranorin and salazinic acid).

Habitats: On oaks and other broadleaf trees, rarely conifers or on rocks, in North Coastal Forest and Valley and Foothill Woodland from near sea level to 2000 ft elevation.

Range: Rather rare from San Luis Obispo County northward to Mendocino County in the North and South Coast Ranges.

Peltigera Willd.

Peltigera is a conspicuous and widespread soil-inhabiting lichen in California. It is recognized in the field by the broad, brown-tinged lobes and the raised or dark veins on the lower surface, visible without a hand lens (see fig. 10c). Apothecia, when produced, are large, often erect, with colorless, three–six septate spores. There is no other lichen genus in California which could be confused with it. Unfortunately, the taxonomy of *Peltigera* is poorly known at present. Color and distinctness of veins, type of rhizines, presence of pruina on the lobe surface, and development of lobules are all rather difficult intergrading characters to differentiate without much guidance and experience.

There are at least 10 species in the state. They are most common in the northern coniferous forests, less well developed south of the San Francisco Bay area, and relatively rare in the

High Sierra where they seem to be smothered out by the heavy snows.

1. Surface of thallus with small raised spots (cephalodia, fig. 12g) visible without hand lens; thallus bright green when wet.
  2. Lower surface with indistinct, flattened veins. . . . . . .
  . . . . *P. aphthosa*
  [2. Lower surface with dark, raised veins. . . . .
  . . . . *P. leucophlebia*]
1. Surface lacking cephalodia; thallus remaining brown when wet (except *P. venosa*).
  3. Thallus sorediate on the surface or margins.
    4. Soralia marginal; collected at base of trees. . . . . .
    . . . . *P. collina*
    [4. Soredia laminal; collected on soil. . . *P. didactyla*]
  3. Thallus lacking soredia.
    5. Thallus 1–2 cm wide, fan-shaped, dark green. . . .
    . . . . *P. venosa*
    5. Thallus larger, 3–12 cm broad, not fan-shaped, brown or whitish brown.
      6. Upper surface shiny to the margin (rarely pruinose at the tips); veins black, not raised. . . . . .
      . . . . *P. polydactyla*
      6. Upper surface dull, covered with a thin tomentum (especially toward the lobe margins); veins pale to brown, raised.
        7. Margins of lobes and surface cracks lobulate.
        . . . . *P. praetextata*
        7. Margins of lobes smooth.
          [8. Lobes crowded, less than 10–15 mm wide, leathery, usually turned up at the tips. . . . . *P. rufescens*]
          8. Lobes expanded, wider, thin, flat without upturned tips.
            9. Thallus firm with pale veins. . . . . .
            . . . . *P. canina*
            [9. Thallus thin and rather fragile, the veins darkening. . . . .
            . . . . *P. membranacea*]

FIG. 38. *Peltigera aphthosa* (×1).

Peltigera aphthosa (L.) Willd. (fig. 38)

Thallus greenish gray, loosely adnate, 8–10 cm broad; lobes 10–20 mm wide, rotund apically, surface with brown warty cephalodia, shiny; lower surface white with light brown, raised veins and tufted rhizines. Apothecia common, erect on lobe tips, 6–7 mm in diameter, disk dark brown. Cortex and medulla K−, C−, P− (tenuiorin).

Habitats: On soil and mosses in Montane Forest and North Coastal Forest from near sea level to 3000 ft elevation.

Range: Rather rare from the Santa Cruz Mountains northward to Del Norte, Siskiyou, and Modoc counties in the North Coast Ranges, Cascades, Klamath Mountains, and Modoc Plateau and in Plumas and Sierra counties on the western slopes of the Sierra Nevada.

Closely related *P. leucophlebia* (Nyl.) Gyel. is distinguished by more distinct veins, but the two species often intergrade.

Peltigera canina (L.) **Willd.** (pl. 8a)

Thallus brownish gray, loosely adnate, 5–10 cm broad; lobes 10–16 mm wide, becoming suberect, surface dull with a fine tomentum; lower surface whitish with raised pale veins, rhizines tufted, tan, sparse. Apothecia common, erect on lobe tips, 3–8 mm in diameter, disk dark brown. Cortex and medulla K−, C−, P− (no substances present).

Habitats: On soil, mosses, and base of trees in North Coastal and Montane Forest and Valley and Foothill Woodland from sea level to 5000 ft elevation.

Range: Rare in Los Angeles County and the San Bernardino Mountains, more common from the Santa Cruz Mountains northward into Oregon in the North Coast Ranges and Klamath Mountains and from Inyo County to Plumas County in the Sierra Nevada.

Peltigera canina represents a constellation of species which are poorly known at this time. For example, *P. membranacea* (Ach.) Nyl. has broad, thin lobes but intergrades over a wide range with *P. canina.* Another very common population on exposed soil with lobes curled up and richly branched rhizines is recognized as *P. rufescens* (Weis) Humb., but many lichenologists do not try to separate it from *P. canina.*

Peltigera collina (Ach.) **Schrad.** (fig. 39a)

Thallus brownish gray, loosely adnate, 3–10 cm broad; lobes 2–5 mm wide, margins curling upward, sorediate, soralia coarse and sometimes appearing isidiate, surface scabrid pruinose toward the tips; lower surface tan, veins flattened, dark, rhizines, tufted, long and branched. Apothecia rarely seen, erect. Cortex and medulla K−, C−, P− (tenuiorin and zeorin).

Habitats: On soil, mosses, humus over boulders, and especially the base of oaks and other broadleaf trees in North Coastal and Montane Forest and Valley and Foothill Woodland from near sea level to 5500 ft elevation.

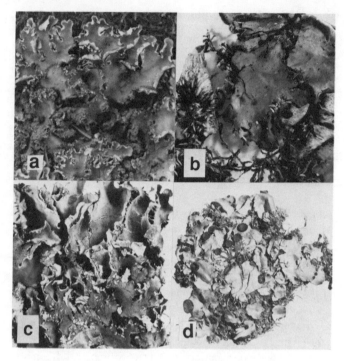

FIG. 39. Species of *Peltigera*: (*a*) *P. collina* (×1); (*b*) *P. polydactyla* (×1); (*c*) *P. praetextata* (×1); (*d*) *P. venosa* (×2).

Range: See fig. 77b.

This is the only consistently corticolous species in the genus, easily recognized by the marginal soredia. Another sorediate species, *P. didactyla* (With.) Laundon (*P. spuria*), has large laminal soralia and occurs on open soil.

Peltigera polydactyla (Neck.) Hoffm.   (fig. 39b)

Thallus greenish or grayish brown, adnate, 6–12 cm broad; lobes fairly long, 10–17 mm wide, semierect, the surface smooth and shiny; lower surface white to buff, with fairly conspicuous dark brown veins, rhizines light brown, sparse. Apothecia common, erect on lobe tips, 3–6 mm broad, disk dark brown. Cortex and medulla K−, C−, P− (tenuiorin, zeorin, and other substances present).

Habitats: On soil, mosses, fallen logs, or base of trees in Montane Forest and North Coastal Forest from near sea level to 3000 ft elevation.

Range: Rather rare from the Santa Cruz Mountains northward in the North Coast Ranges and Klamath Mountains and in Plumas County on the western slopes of the Sierra Nevada.

Peltigera praetextata (Somm.) Zopf (fig. 39c)

Thallus dark or brownish gray, loosely adnate, 6–11 cm broad; lobes rather long, 8–10 mm wide, the surface dull, in part covered with isidia-like regeneration lobules along the margins and cracks; lower surface with raised light brown veins, rhizines tufted, conspicuous. Apothecia lacking. Cortex and medulla K−, C−, P− (no substances).

Habitats: On soil or mosses in North Coastal and Montane Forest and Valley and Foothill Woodland from 2000 to 5000 ft elevation.

Range: Rare in the San Bernardino Mountains, but fairly common from the Santa Cruz Mountains to Siskiyou County in the Coast Ranges and Klamath Mountains and from Placer County to Shasta County in the Sierra Nevada.

Peltigera venosa (L.) Hoffm. (fig. 39d)

Thallus brownish green (bright green when wet), adnate on soil, to 2 cm broad, fan-shaped; surface smooth; lower surface white with dark brown veins, sparsely rhizinate. Apothecia common, marginal and horizontal, 2–3 mm wide, disk black. Cortex and medulla K−, C−, P− (tenuiorin, zeorin, and unidentified substances).

Habitats: On soil banks along trails in Montane Forest at 500 to 3000 ft elevation.

Range: Rare from the Santa Cruz Mountains northward to Del Norte County in the North Coast Ranges and Cascades and from Mariposa County to Plumas County on the western slopes of the Sierra Nevada.

Phaeophyscia Moberg

This genus is a recent segregate from *Physcia* (see Esslinger, 1978). The main characters are the small dark greenish gray thallus reacting K− (atranorin lacking), black (rarely pale) lower surface, and small ellipsoidal conidia. The spores, as in *Physcia,* are brown and two-celled.

Phaeophyscia is widespread in California, but it has not been collected frequently, perhaps because it is rather inconspicuous. The corticolous species prefer the base of oak trees in the South Coast Ranges. The fragile saxicolous species are often very difficult to collect intact.

1. Thallus lacking soredia; apothecia often present.
    2. Collected on rocks. . . . . *P. decolor*
    [2. Collected on trees. . . . . *P. ciliata*]
1. Thallus sorediate; apothecia rare or absent.
    3. Soredia produced mostly in labriform soralia. . . . .
        . . . . *P. hirsuta*
    3. Soredia produced in marginal or laminal soralia.
        4. Tips of lobes with fine hairs (use hand lens); soredia marginal. . . . . *P. cernohorskyi*
        4. Hairs lacking; soralia mostly laminal.
            5. Lower surface distinct, black (use hand lens).
                . . . . *P. orbicularis*
            [5. Lower cortex lacking, the lobes tightly adnate.
                . . . . *Hyperphyscia adglutinata*]

***Phaeophyscia cernohorskyi* (Nadv.) Essl.**   (fig. 40a)

Thallus dark gray, closely adnate, 1−2 cm broad; lobes short and narrow, about 0.5 mm wide, the surface with fine white hairs toward the margin (use hand lens), sorediate marginally; lower surface variable, tan or darker, sparsely rhizinate. Apothecia rare, rim of the disk with white hairs. Cortex and medulla K−, C−, P− (no substances present).

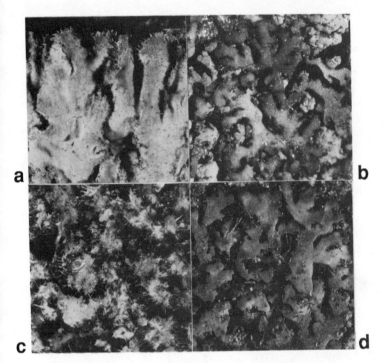

FIG. 40. Species of *Phaeophyscia:* (*a*) *P. cernohorskyi;* (*b*) *P. decolor;* (*c*) *P. hirsuta;* (*d*) *P. orbicularis* (all ×10).

Habitats: On Valley Oak and Coast Live Oak, rarely on rocks, in the Valley and Foothill Woodland from near sea level to 3000 ft elevation.

Range: Widespread from San Diego County northward to San Luis Obispo County.

Phaeophyscia decolor (Kashiwadani) Essl.   (fig. 40b)

Thallus dark to brownish gray or faintly white pruinose, closely adnate, 3–4 cm broad; lobes narrow, 0.5–1 mm wide; lower surface black, sparsely rhizinate. Apothecia common, about 1 mm in diameter. Cortex and medulla, K−, C−, P− (no substances present).

Habitats: On sheltered granitic rocks in Montane Forest at 3000 to 7000 ft elevation.

Range: Rare (but probably overlooked) in Mariposa County and Alpine County on the western slopes of the Sierra Nevada.

A close relative on trees is *P. ciliata* (Hoffm.) Moberg, which has been collected in Mariposa County.

Phaeophyscia hirsuta (Meresch.) Essl.   (fig. 40c)

Thallus brownish mineral gray, closely adnate, 2–4 cm broad; lobes short and narrow, 0.5–1 mm wide, sorediate on margin and lower surface of tips, the soralia labriform, turning upward, short translucent hairs forming on the soredia and cortex; lower surface black, densely rhizinate. Apothecia rare, about 1 mm in diameter with hairs around the rim. Cortex and medulla K−, C−, P− (no substances present).

Habitats: On Valley Oak and Interior Live Oak in Valley and Foothill Woodland from sea level to 1000 ft elevation.

Range: Rare in Santa Clara and Los Angeles counties in the South Coast Ranges.

Phaeophyscia orbicularis (Neck.) Essl.   (fig. 40d)

Thallus greenish to brownish gray, adnate, 1–3 cm broad; lobes narrow, 1–2 mm wide, sorediate marginally, the soralia orbicular; lower surface brown to black. Cortex and medulla K−, C−, P− (no substances present).

Habitats: On oaks and other broadleaf trees, rarely on rocks, in Valley and Foothill Woodland and Montane Forest from 500 to 5500 ft elevation.

Range: See fig. 77c.

When this species assumes a pale greenish color, it may be confused with *Physcia,* which would react quickly K+ yellow on the surface. *Hyperphyscia adglutinata* is also close, having a K− reaction, but lacks rhizines and is more tightly appressed on the bark.

Physcia (Schreb.) Michaux

Physcia is a small, narrow-lobed, closely adnate lichen. Many species have a characteristic whitish cast because of a fine pruina which develops on the upper cortex. All have a white to pale tan lower surface with pale rhizines and react K+ yellow in the cortex (atranorin). Some species also react yellow in the medulla and contain both atranorin and zeorin. Two externally similar species of *Heterodermia* in California lack a lower cortex, and three closely related genera, *Phaeophyscia, Hyperphyscia,* and *Physconia,* formerly classified under *Physcia,* lack atranorin (cortex K−). The spores are normal for the Physciaceae, brown and two-celled.

The position of soralia and presence of white spotting in the cortex are important characters for species identification. It is also extremely important to determine the color reaction in the exposed medulla (K+ yellow or K−) accurately using a hand lens or binocular dissecting microscope. The taxonomy of the western species, however, is not well known at this time; some names now in use may be incorrect and some species undescribed. Do not be discouraged if some specimens do not fit the keys and descriptions well.

Physcia is one of the most widespread lichens in the Valley and Foothill Woodland, and virtually every oak tree has one or more species. At higher elevations one finds a number of other species growing on sheltered faces of large outcrops, very brittle when dry and hard to collect.

1. Thallus sorediate; apothecia rarely observed.
    2. Marginal cilia present.
        3. Tips of lobes inflated and hood-shaped with soredia under hoods. .... *P. adscendens*
        3. Tips of lobes flat, not inflated; soredia on lower surface of lobe tips. .... *P. tenella*
    2. Cilia lacking (some rhizines may project from below).
        4. Soralia laminal, capitate; surface more or less distinctly white-spotted (see fig. 12f); medulla K+ yellow. .... *P. caesia*
        4. Soralia marginal, on the lower surface, or diffuse; white spots lacking; medulla K−.

5. Soredia on lower surface of lobe tips; lobes very short and narrow, less than 1 mm wide. . . . . . .
. . . . *P. dubia*
5. Soredia on margins or surface of lobes; lobes longer and branched, 1–2 mm wide.
  6. Soredia granular on dissected lobe tips; collected on rocks and trees. . . . . . . *P. callosa*
  [6. Soredia densely produced over the upper surface; collected only on trees. . . . .
. . . . *P. clementei*]
1. Thallus lacking soredia; apothecia almost always present.
  7. Collected on trees.
    8. Medulla K+ yellow; white spotting of cortex conspicuously developed (see fig. 12f). . . . *P. aipolia*
    8. Medulla K−; white spotting absent or weakly developed.
      9. Lobe surface shiny, lacking white pruina. . . . . .
. . . . *P. stellaris*
      9. Lobes becoming white pruinose. . . . . *P. biziana*
  7. Collected on rocks.
    10. Lobe surface white spotted; medulla K+ yellow.
. . . . *P. phaea*
    [10. White spotting absent; medulla K−. . . . .
. . . . *P. albinea*]

Physcia adscendens (Fr.) Oliv.   (fig. 41a)

Thallus light gray, adnate on bark but with ascending lobe tips, 1–3 cm broad; lobes 0.5–1 mm wide, linear, soredia present under hooded lobe tips, the margins short ciliate; lower surface white, moderately rhizinate. Apothecia lacking. Cortex K+ yellow; medulla K−, C−, P− (atranorin).

Habitats: On oaks and other broadleaf trees, conifers, and rarely rocks in North Coastal and Montane Forest and Valley and Foothill Woodland from 500 to 6000 ft elevation.

Range: See fig. 77d.

This unique hooded species typically grows in small colonies on twigs. When the hoods are only weakly inflated, there

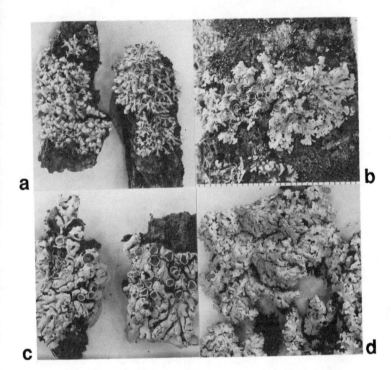

FIG. 41. Species of *Physcia:* (a) *P. adscendens;* (b) *P. aipolia;* (c) *P. biziana;* (d) *P. caesia* (all ×1).

is possible confusion with *P. tenella,* which has flattened lobe tips.

Physcia aipolia (Ehrh.) Fürnrohr (fig. 41b)

Thallus whitish gray, closely adnate, 2–4 cm broad; lobes 1–2 mm wide, linear but crowded, surface white-spotted (under hand lens); lower surface white, moderately rhizinate, rhizines white with darkening tips. Apothecia very common, 0.5–1.5 mm wide, disk black or becoming heavily white pruinose. Cortex and medulla K+, P+ yellow, C− (atranorin and zeorin).

Habitats: On oaks and other broadleaf trees and conifers in

Valley and Foothill Woodland and Montane Forest from 1000 to 4500 ft elevation.

Range: See fig. 78a.

Physcia biziana (Mass.) Zahlbr. (fig. 41c)

Thallus whitish gray, closely adnate, 2–5 cm broad; lobes 1–2 mm wide, the surface lightly pruinose; lower surface white to tan, moderately rhizinate. Apothecia common, 1–2 mm wide, disk black or white pruinose. Cortex K+ yellow; medulla K–, C–, P– (atranorin).

Habitats: On oaks and other broadleaf trees in Valley and Foothill Woodland from 500 to 5500 ft elevation.

Range: Fairly common from San Diego County northward to Colusa County in the North and South Coast Ranges and from Kern County to Calaveras County in the western front of the Sierra Nevada.

The amount of pruina varies from very dense over the whole thallus to just toward lobe tips. There is obvious intergradation with *P. stellaris,* which is on the average smaller and more delicate.

Physcia caesia (Hoffm.) Fürnrohr (fig. 41d)

Thallus whitish gray, closely adnate to adnate, 2–3 cm broad; lobes 0.5–1.5 mm wide, linear, the surface white-spotted, soralia laminal in capitate groups; lower surface white to tan, moderately rhizinate. Apothecia lacking. Cortex and medulla K+, P+ yellow, C– (atranorin and zeorin).

Habitats: On rocks in Montane and Subalpine Forest from 1000 to 8000 ft elevation.

Range: Rather rare from Riverside County to Los Angeles and in the Santa Cruz Mountains in the South Coast Ranges and from Tulare and Inyo counties to Siskiyou County in the Sierra Nevada, Modoc Plateau, and Cascades.

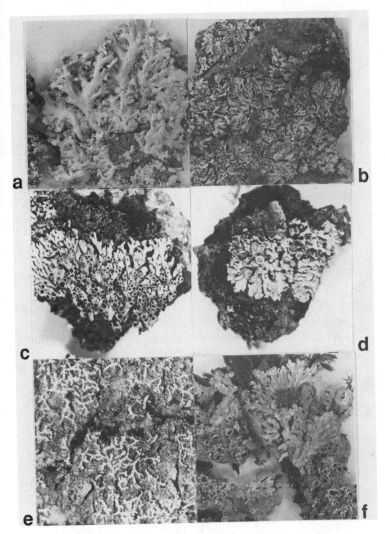

FIG. 42. Species of *Physcia* and *Physconia*: (a) *Physcia callosa*; (b) *P. dubia*; (c) *P. phaea*; (d) *P. stellaris*; (e) *P. tenella*; (f) *Physconia distorta* (all about ×1).

Physcia callosa Nyl.   (fig. 42a)

Thallus whitish gray, adnate, 2–4 cm broad, often coalescing into large colonies; lobes rather short, 1–1.5 mm wide, apically dissected with scattered marginal soredia; lower surface white, moderately rhizinate. Apothecia lacking. Cortex K+ yellow; medulla K−, C−, P− (atranorin).

Habitats: On rocks, rarely broadleaf trees and conifers, in Valley and Foothill Woodland from near sea level to 6000 ft elevation.

Range: Fairly common from San Diego County northward to Sonoma County in the North and South Coast Ranges and rare on the western slopes of the Sierra Nevada from Mariposa County to Butte County.

This is an extremely variable species. It is most typical on rocks in the Coast Ranges; tree forms may have greater development of soredia. A related species, *P. clementei* (Turn.) Lynge, is sorediate over the whole center of the thallus; it has been collected from western Kern County south to Los Angeles County in the South Coast Ranges.

Physcia dubia (Hoffm.) Lett.   (fig. 42b)

Thallus whitish gray, closely adnate on rock, 1–3 cm broad; lobes short and little branched, 0.5–1.0 mm wide, the tips becoming somewhat dissected, sorediate on the margin and lower surface of lobe tips; lower surface white, moderately rhizinate. Apothecia not seen. Cortex K+ yellow; K−, C−, P− (atranorin).

Habitats: On rocks in pastures and open areas in Valley and Foothill Woodland, Montane Forest, and Subalpine Forest from 1000 to 9000 ft elevation.

Range: Rather common in the San Bernardino Mountains, in the South Coast Ranges, in Trinity and Siskiyou counties in the Klamath Mountains, and from Kern County to Lassen County in the western foothills of the Sierra Nevada and Modoc Plateau.

Physcia phaea (Tuck.) Thoms.   (fig. 42c)

Thallus whitish gray, closely adnate, 2–6 cm broad; lobes linear, 0.5–1 mm wide, the surface white-spotted; lower surface white, moderately rhizinate. Apothecia very common, 0.5–1.5 mm wide, disk black. Cortex and medulla K+, P+ yellow, C– (atranorin and zeorin).

Habitats: On rocks in sheltered areas in Valley and Foothill Woodland and Montane Forest from near sea level to 6000 ft elevation.

Range: Rather rare from San Diego County northward to Mt. Diablo in the San Francisco Bay area and Glenn County in the North and South Coast Ranges and from Kern County to Modoc County on the western slopes of the Sierra Nevada and Modoc Plateau.

Physcia stellaris (L.) Nyl.   (fig. 42d)

Thallus whitish gray, closely adnate, 2–4 cm broad; lobes short, 0.5–2 mm wide, surface shiny, margins somewhat revolute; lower surface white, moderately rhizinate, rhizines clustered, white with darkening tips. Apothecia very common, 0.6–3 mm wide, disk black or becoming heavily white pruinose. Cortex K+ yellow; medulla K–, C–, P– (atranorin).

Habitats: On oaks and other broadleaf trees and conifers in Montane Forest, Valley and Foothill Woodland, and Coastal Scrub from near sea level to 5200 ft elevation.

Range: See fig. 78b.

This is a common lichen on twigs of oak trees. There is considerable variation in size. *Physcia biziana* is very similar but is uniformly covered with white frosty pruina. The two species obviously intergrade and are difficult to tell apart at times. Another similar K– species, *P. albinea* (Ach.) Nyl., has longer, narrow lobes and grows on rock at higher elevations.

Physcia tenella (Scop.) DC.   (fig. 42e)

Thallus whitish gray, closely adnate, about 1 cm wide; lobes narrow, 0.3–1.0 mm wide, the tips subascending; surface white-spotted, the margins ciliate, cilia up to 1 mm long, sorediate, the soredia sparsely produced under lobe tips which remain flattened; lower surface white with sparse rhizines. Apothecia not seen. Cortex K+ yellow; medulla K–, C–, P– (atranorin).

Habitats: On oaks and other broadleaf trees and shrubs in North Coastal and Montane Forest and Valley and Foothill Woodland from near sea level to 6800 ft elevation.

Range: Widespread from Riverside County to Mendocino County in the North and South Coast Ranges and from Kern County to Shasta County on the western slopes of the Sierra Nevada.

The sorediate tips are flattened and the cilia conspicuous. *Physcia adscendens* has soredia in hood-shaped inflated lobe tips, but intergradations with *P. tenella* often cause trouble. *Heterodermia leucomelaena* with short lobes might also be keyed here, but the lower surface is cottony and lacks a cortex.

Physconia Poelt

Physconia is a narrow-lobed, adnate, foliose genus formerly classified as *Physcia*. It differs from *Physcia* in lacking atranorin (cortex K–) and in having different, larger spores and a more complex cortical structure. Two of the species in California, *P. detersa* and *P. distorta* (*P. pulverulenta* in older works), are among the most commonly collected lichens in the oak woodlands.

All of the species vary tremendously in color, from greenish brown to pure white, depending on the amount of white pruina developed. This variability may cause problems in identification because of apparent intergradation with smaller *Phaeophyscia orbicularis* (laminal soralia) on the one hand or with *Physcia* species, such as large specimens of heavily white pruinose *P. biziana* (cortex K+ yellow), on the other. We should also mention that the taxonomy of *Physconia* in western North

America is poorly known and name changes are bound to occur when the species are revised.

1. Soredia present on or under lobe margins; apothecia rarely seen.
   2. Soredia whitish. . . . . *P. detersa*
   [2. Soredia with a yellowish cast. . . . . *P. enteroxantha*]
1. Soredia lacking; lobulate apothecia common.
   3. Growing on trees; white pruina highly developed. . . .
. . . . *P. distorta*
   [3. Growing on soil and mosses in Alpine Fell-Field; pruina not so conspicuous. . . . . *P. muscigena*]

Physconia detersa (Nyl.) Poelt   (fig. 12e, pl. 8b)

Thallus whitish gray to brownish, adnate, 2–4 cm broad; lobes 0.5–1 mm wide, marginally sorediate, the surface becoming heavily white pruinose; lower surface black, densely rhizinate with squarrose rhizines. Apothecia rare, 1 mm wide, disk black or white pruinose. Cortex and medulla K−, C−, P− (no substances present).

Habitats: On oak and other broadleaf trees, rarely on rock or mosses over rock, in Valley and Foothill Woodland and Montane Forest from 1000 to 6800 ft elevation.

Range: See fig. 78c.
   This is the most frequently collected lichen in California, occurring mostly on oak. A form with soredia on the lower surface of lobe tips may be found from Sutter County southward in the Sierra Nevada. Rarer *P. enteroxantha* (Nyl.) Poelt is barely distinguishable from *P. detersa* except for the yellowish soredia.

Physconia distorta (With.) Laundon   (fig. 42f)

Thallus brownish to whitish gray, adnate, 2–5 cm broad; lobes linear, 0.5–1 mm wide, the surface heavily pruinose, margins often lobulate; lower surface black, moderately rhizinate, the rhizines squarrose, clumped. Apothecia very common, 0.5–3 mm wide, disk black to pruinose, rim often lobulate. Cortex and medulla K−, C−, P− (no substances present).

Habitats: On oaks and other broadleaf trees and juniper in Valley and Foothill Woodland and Montane Forest from 500 to 5000 ft elevation.

Range: See fig. 78d.

This common lichen usually has conspicuously lobulate apothecia and dense white pruina. Closely related *P. muscigena* is a rare alpine lichen in the High Sierra. It is usually dark greenish brown and grows on soil or humus on rocks.

Platismatia Culb. & Culb.

This conspicuous genus has large, suberect lobes, white angular pseudocyphellae, and a bare lower surface. Apothecia when present are marginal and have simple, colorless spores. The four species in California were placed in *Cetraria* by older workers. They are all found typically on conifers, less commonly on deciduous trees. *Platismatia glauca* is quite common in the northern half of California, whereas *P. herrei* and *P. stenophylla* are strictly limited to the North Coast Ranges. The fourth species, *P. lacunosa,* is known rarely in Humboldt County; it lacks soredia and isidia and has a large, deeply ridged, foveolate thallus.

1. Lobe margins isidiate, isidiate-lobulate, or sorediate.
2. Lobes broad, 3–10 mm wide, predominantly sorediate. . . . . *P. glauca*
2. Lobes narrow and linear, 2–4 mm wide; mostly isidiate. . . . . *P. herrei*
1. Margins of lobes lacking soredia and isidia.
3. Lobes narrow and linear, 2–4 mm wide; apothecia often present. . . . . *P. stenophylla*
[3. Lobes broad, 3–10 mm wide, deeply foveolate; apothecia rare. . . . . *P. lacunosa*]

Platismatia glauca (L.) Culb. & Culb. (pl. 8c)

Thallus mineral gray to gray-green, loosely attached to suberect, 3–11 cm broad; lobes variable in width, 3–5 mm wide, short to rather elongate, surface smooth to slightly wrinkled, margins dissected with granular to subisidiate soredia; lower surface black and wrinkled at the center, becoming mottled brown and white and smooth at the margins. Apothecia lacking. Cortex K+ yellow; medulla K−, C−, P− (atranorin and fatty acids).

Habitats: On conifers, Madrone, Tanbark Oak, and other broadleaf trees in North Coastal and Montane Forests and higher-elevation Valley and Foothill Woodland from sea level to 5500 ft elevation.

Range: See fig. 79a.

Platismatia herrei (Imsh.) Culb. & Culb. (fig. 43a)

Thallus light gray-green, loosely attached to suberect, 2−8 cm broad; lobes 1−3 mm wide, strap-shaped, surface shiny and white-spotted, smooth to slightly wrinkled with a few isidia, margins finely dissected with dense branched isidia; lower surface variable, black to brown with mottled white, smooth to somewhat wrinkled. Apothecia lacking. Cortex K+ yellow; medulla K−, C−, P− (atranorin and fatty acids).

Habitats: On conifers in the North Coastal and Montane Forests from sea level to 3000 ft elevation.

Range: Fairly common from Monterey County northward to Oregon in the North Coast Ranges.

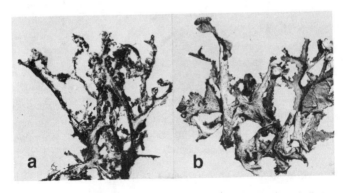

FIG. 43. Species of *Platismatia:* (a) *P. herrei;* (b) *P. stenophylla* (×1).

Platismatia stenophylla (Tuck.) Culb. & Culb.   (fig. 43b)

Thallus light mineral gray to gray-green, loosely attached to suberect, 5–10 cm broad; lobes 3–5 mm wide, strap-shaped, surface smooth to wrinkled, the margins with black pycnidia; lower surface white to mottled brown, smooth to wrinkled. Apothecia rather rare, terminal, 1–9 mm wide, disk light brown. Cortex K+ yellow; medulla K−, C−, P− (atranorin and fatty acids).

Habitats: On conifers in North Coastal Forest from 500 to 3500 ft elevation.

Range: Rather rare from the Santa Cruz Mountains northward into Oregon in the North Coast Ranges.

Pseudocyphellaria Vain.

This is a small but conspicuous genus of three species in California. The broad foveolate lobes are brown when dry and dullish green-brown when wet. Apothecia are large and contain colorless, transversely septate spores. The most characteristic feature, which separates the genus from all others, is the lower surface—finely brown tomentose with numerous tiny white pores (yellow in *P. crocata*) visible without a hand lens. These pores are the pseudocyphellae. The genus usually grows on hardwood trees and is fairly common from just south of the San Francisco Bay area to Del Norte County in the North Coast Ranges and the Klamath Mountains but much rarer in the Sierra Nevada foothills from Calaveras County to Plumas County. The species are easily distinguished by the presence of apothecia or soredia and color of the medulla.

1. Thallus with numerous apothecia; soredia absent. . . . .
. . . . *P. anthraspis*
1. Thallus rarely with apothecia but always sorediate.
   2. Soredia and pores white or darkening. . . . *P. anomala*
   [2. Soredia and pores yellow. . . . . *P. crocata*]

Pseudocyphellaria anomala Ahti & Brodo   (pl. 8d)

Thallus brown to grayish brown, loosely adnate, large, 4–9 cm broad; lobes 6–14 mm wide, little branched, the surface strongly ridged with gray-black soredia developing along the margins and ridges; lower surface tan with short tomentum and numerous white pseudocyphellae. Apothecia rare, 1.5–2 mm wide, disk dark brown. Cortex K−; medulla K+ yellow, P+ orange (stictic and constictic acids and unidentified terpenes).

Habitats: On Valley Oak, California Black Oak, and other broadleaf trees in North Coastal and Montane Forest and Valley and Foothill Woodland from 500 to 4500 ft elevation.

Range: See fig. 79b.

There is only one very similar species with raised white "pores" below, *Nephroma resupinatum,* which has apothecia on the undersurface of lobe tips. The third species of *Pseudocyphellaria* in California, *P. crocata* (L.) Vain., is unmistakable: It has brilliant yellow soredia and pores. It is, however, very rare, known only on sheltered rocks in oak pastures in Humboldt County and in the Santa Cruz Mountains.

Pseudocyphellaria anthraspis (Ach.) Magn. (fig. 44)

Thallus light brown to brown, loosely adnate, large, 7–14 cm broad; lobes 7–15 mm wide, the upper surface strongly ridged; lower surface tan, short tomentose with numerous white pseudocyphellae. Apothecia common along the ridges, 1.5–3 mm

FIG. 44. *Pseudocyphellaria anthraspis* (×1).

wide, disk brown to dark brown. Cortex K−; medulla K+ yellow and P+ orange (stictic and constictic acids with unidentified terpenes).

Habitats: On Valley Oak, California Black Oak, and other broadleaf trees or on rocks in North Coastal and Montane Forest and Valley and Foothill Woodland from near sea level to 4500 ft elevation.

Range: See fig. 79c.

Punctelia Krog

Punctelia is characterized by round white pores (pseudocyphellae) on the upper surface of the thallus (see fig. 12b), visible with the naked eye. The species included here were formerly classified in *Parmelia,* which has angular pseudocyphellae. Both genera have the same apothecial characters with simple, colorless spores. The three species of *Punctelia* in California occur on trees along roadsides and in open grazing land at lower elevations, especially in the Coast Ranges.

1. Lower surface pale brown. . . . . *P. subrudecta*
1. Lower surface black.
    2. Thallus distinctly brownish, always occurring on rocks.
    . . . . *P. stictica*
    [2. Thallus pale greenish gray, occurring on trees. . . . . . .
    . . . . *P. borreri*]

Punctelia stictica (Del.) Krog   (fig. 45a)

Thallus brownish, closely adnate, 3–5 cm broad; lobes 2–3 mm wide, crowded, shiny, wrinkled toward the center with prominent white pseudocyphellae, becoming sparsely sorediate from the pores; lower surface black, becoming brown at the margins, rhizines sparse. Apothecia lacking. Cortex K+ yellow; medulla K−, C+, KC+ rose, P− (atranorin and gyrophoric acids).

Habitats: On rocks in open pastures in North Coastal Forest from near sea level to 1200 ft elevation.

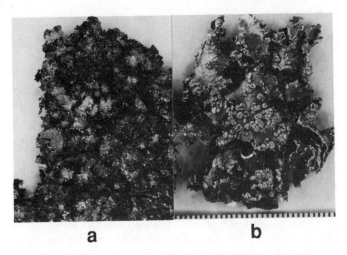

FIG. 45. Species of *Punctelia:* (a) *P. stictica;* (b) *P. subrudecta* (×1).

Range: Locally abundant in the Santa Cruz Mountains and in Humboldt County in the North Coast Ranges.

Punctelia subrudecta (Nyl.) Krog   (fig. 45b)

Thallus greenish gray, adnate to loosely adnate, 5–8 cm broad; lobes irregularly broadened, 3–5 mm wide, the surface pseudocyphellate, sorediate laminally and marginally; lower surface tan, rhizinate nearly to the margin. Apothecia lacking. Cortex K+ yellow; medulla K−, C+, P− (atranorin and lecanoric acid).

Habitats: On oaks and other broadleaf trees in North Coastal Forest and Valley and Foothill Woodland from near sea level to 2800 ft elevation.

Range: See fig. 79d.
   A closely related gray species with a black lower surface and gyrophoric acid, *P. borreri* (Sm.) Krog, has been collected in Santa Barbara County on oak trees. Except for the black lower surface it is very similar to *P. subrudecta*.

Rhizoplaca Zopf

This is a closely adnate, umbilicate genus formerly classified in *Lecanora*. The apothecia are usually conspicuously developed and either flesh-colored or greenish with simple colorless spores. It is easily told from *Umbilicaria* by the yellow-green thallus color. The two species known in California often grow side by side on large, exposed outcrops.

1. Apothecial disk pale yellowish to flesh-colored. . . . .
. . . . *R. chrysoleuca*
[1. Apothecial disk darker, greenish. . . . *R. melanophthalma*]

Rhizoplaca chrysoleuca (Sm.) Zopf   (pl. 9d)

Thallus yellow-green, adnate but easily plucked off with a knife, 1–3 cm broad; lobes 1–3 mm wide, crowded and fused, the margins black; lower surface light brown, rhizines lacking. Apothecia common, 1–3 mm wide, disk flesh-colored to pale orange; spores simple, colorless, 6 × 12 μm. Cortex K−; medulla K−, C−, P− (usnic and pseudoplacodiolic acids).

Habitats: On exposed acidic rocks in Montane and Subalpine Forest from 4000 to 8000 ft elevation.

Range: Widespread and locally abundant in Riverside County in the San Bernardino Mountains, in Siskiyou and Modoc counties in the Sierra Nevada, and in Modoc County in the Modoc Plateau.

A second species in the genus, *R. melanophthalma* (DC.) Leuck. & Poelt, is also common on exposed rocks and often grows with *R. chrysoleuca*. It differs in having a dark gray disk and a brown to blackening lower surface.

Sticta (Schreb.) DC.

Sticta, a broad-lobed brownish lichen, can be recognized instantly by the large white pores (cyphellae) scattered in the tomentum on the lower side (see fig. 12a). The phycobiont is blue-green. There are two rather rare species in California in the moist North Coast Ranges and Klamath Mountains, growing at the base of trees or over mosses on rocks or trees.

[1. Pores raised; apothecia on lower surface of lobe tips. . . .
. . . . *Nephroma resupinatum*]
  1. Pores sunken; apothecia (if present) laminal.
     2. Margins of lobes sorediate. . . . . *S. limbata*
     2. Surface of lobes isidiate. . . . . *S. fuliginosa*

Sticta fuliginosa (Hoffm.) Ach.   (fig. 46a)

Thallus dark greenish brown, loosely adnate, 3–10 cm broad; lobes 8–15 mm wide, little branched, the surface dull, uniformly covered with short, round isidia; lower surface tan with long tomentum, cyphellae scattered in the tomentum. Apothecia lacking. Cortex and medulla K–, C–, P– (no substances present).

Habitats: On base of trees or over mosses on trees or rocks in North Coastal Forest from near sea level to 3000 ft elevation.

Range: Rather rare from the Santa Cruz Mountains northward to Oregon in the North Coast Ranges.

Sticta limbata (Sm.) Ach.   (fig. 46b)

Thallus light brown to brown, loosely adnate, 3–6 cm broad; lobes little branched, 6–11 mm wide, the surface smooth with

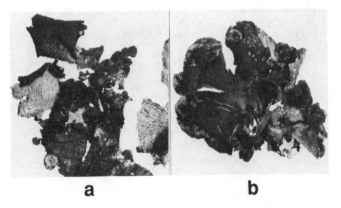

FIG. 46. Species of *Sticta:* (a) *S. fuliginosa;* (b) *S. limbata* (×1).

a few cracks, sorediate on the surface, especially along the cracks, the soredia gray; lower surface tan, short tomentose, cyphellae small, sparse. Apothecia lacking. Cortex and medulla K−, C−, P− (no substances present).

Habitats: On broadleaf trees, rarely conifers, or over mosses on trees in North Coastal Forest from 1000 to 5000 ft elevation.

Range: Rare from the Santa Cruz Mountains northward into Oregon in the North Coast Ranges.

Tuckermannopsis Gyel.

This small-lobed, foliose genus is often classified with *Cetraria*. The true Cetrarias, however, are now considered to center around *C. islandica* (Iceland Moss) and other species with erect, free-growing lobes, none of which occurs in California. The most important characters used to recognize *Tuckermannopsis* are the marginal apothecia and erect or immersed marginal pycnidia (see fig. 13), as opposed to laminal apothecia and pycnidia in related parmelioid genera. Thallus color varies widely from the golden lemon yellow of *T. canadensis* to the brown of *T. chlorophylla* and *T. orbata*. The lower surface is pale tan or brown with sparse rhizines. The apothecia have colorless simple spores. The seven species in California are separated by thallus color, soredia, cilia, and presence of papillae. They are typically found on conifers in Montane Forest.

1. Thallus pale yellow to lemon yellow.
    2. Medulla and lower surface yellow. . . . . . *T. canadensis*
    2. Medulla white, lower surface pale tan. . . . *T. pallidula*
1. Thallus whitish gray, brown, or greenish black.
    3. Thallus whitish or greenish gray.
        [4. Lower surface black, dull, wrinkled; pseudocyphellae lacking. . . . . *Esslingeriana idahoensis*]
        [4. Lower surface mottled white-black, smooth, and shiny; white angular pseudocyphellae present on the surface. . . . . *Platismatia*]
    3. Thallus brown to greenish black.
        5. Powdery soredia present on lobe margins. . . . .
        . . . . *T. chlorophylla*

5. Soredia lacking.
   6. Lobes 3–6 mm broad, the surface becoming strongly papillose. . . . . *T. platyphylla*
   6. Lobes 1–4 mm wide, smooth.
      7. Thallus 3–5 cm broad, adnate to suberect, light brown above with a tan lower surface.
      . . . . *T. orbata*
      7. Thallus 1–3 cm broad, becoming suberect or almost subfruticose, uniformly olive or brownish black.
         8. Lobes short and crowded, flattened; apothecia marginal. . . . . *T. merrillii*
        [8. Lobes becoming elongated, rounded; apothecia terminal. . . . .
        . . . . *Cornicularia californica*]

Tuckermannopsis canadensis (Räs.) Hale   (pl. 1c)

Thallus pale to deep lemon yellow, loosely adnate to suberect on bark, 1.5–4.5 cm broad; lobes short, 2–4 mm wide, surface foveolate, margins with small black pycnidia, becoming dissected; medulla yellow; lower surface yellow, strongly ridged, rhizines sparse or lacking. Apothecia common, 1–4 mm wide, disk brown. Cortex and medulla K−, C−, P− (pinastric, usnic, vulpinic, and fatty acids).

Habitats: On twigs of redwood, Douglas Fir, pines, or rarely broadleaf trees in North Coastal and Montane Forest from 2500 to 4000 ft elevation.

Range: See fig. 80c.

Tuckermannopsis chlorophylla (Willd.) Hale   (pl. 10c)

Thallus brown to olive brown, loosely adnate on bark, 2.5–5 cm broad; lobes rather linear, 1–4 mm wide, margins and in part surface powdery sorediate; lower surface light brown, wrinkled, sparsely rhizinate. Apothecia not seen. Cortex and medulla K−, C−, P− (fatty acids).

Habitats: On base of conifers, fenceposts, fallen logs, and the like in North Coastal and Montane Forest from 1000 to 4500 ft elevation.

Range: Rare in Santa Barbara County in the South Coast Range but becoming common from Santa Cruz County northward to Humboldt and Siskiyou counties in the North Coast Ranges and Klamath Mountains and from Mariposa County to Plumas County on the western slopes of the Sierra Nevada.

Tuckermannopsis merrillii (DR.) Hale (fig. 47a)

Thallus dark greenish to blackish brown, loosely adnate to suberect, 1–2 cm broad; lobes narrow and short, 1–2 mm wide, margins densely papillate with numerous pycnidia; lower surface brown, wrinkled, rhizines sparse. Apothecia common, 1–4 mm wide, disk brown with a papillate rim. Cortex and medulla K−, C−, P− (fatty acids).

Habitats: On twigs or canopy branches of exposed conifers in Montane and Subalpine Forest from 2400 to 8000 ft elevation.

Range: Widespread in San Diego and Santa Barbara counties and Monterey County to Humboldt County in the North and South Coast Ranges and Klamath Mountains and from Mari-

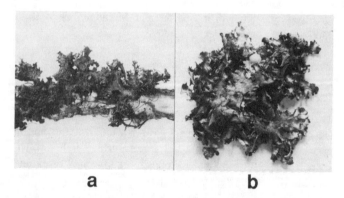

a   b

FIG. 47. Species of *Tuckermannopsis:* (a) *T. merrillii* (×2); (b) *T. orbata* (×1).

posa County to Modoc County in the Sierra Nevada and the Modoc Plateau.

This lichen is unusual in the blackish olive-green color. It intergrades with rare *Cornicularia californica* (Tuck.) DR., which is more distinctly fruticose, but the relationship of the two species is still not well understood.

Tuckermannopsis orbata (Nyl.) Lai (fig. 47b)

Thallus varying in color from light brown to brown, loosely adnate on bark, 3–5 cm broad; lobes short, 2–6 mm wide, surface weakly wrinkled, rhizines sparse. Apothecia common, 3–10 mm wide, disk brown, rim often spinulate. Cortex and medulla K−, C−, P− (fatty acids).

Habitats: On twigs of conifers (especially Douglas Fir) in North Coastal and Montane Forest from sea level to 6500 ft elevation.

Range: Fairly common from Santa Barbara and Santa Cruz counties northward to Del Norte and Siskiyou counties in the North and South Coast Ranges and Klamath Mountains and from Calaveras County to Shasta County on the western slopes of the Sierra Nevada.

This lichen typically grows scattered on twigs with *T. platyphylla*. There is some degree of intergradation between these two: Small or immature specimens of *T. platyphylla* may be misidentified as *T. orbata*, but mature specimens are so much larger and coarsely papillate that there will be no doubt.

Tuckermannopsis pallidula (Tuck.) Hale (fig. 13d)

Thallus light yellow-green, adnate to suberect on bark, 1–2.5 cm broad; lobes somewhat linear, 3–4 mm wide, surface smooth to weakly wrinkled, margins with pycnidia, becoming dissected; lower surface pale, weakly wrinkled, rhizines sparse, tufted. Apothecia common, 2–6 mm wide, disk tan to brown, rim spinulate. Cortex and medulla K−, C−, P− (usnic and fatty acids).

Habitats: On twigs of Douglas Fir and other conifers in Montane Forest from 800 to 4000 ft elevation.

Range: Rare but conspicuous from Calaveras County north-ward to Plumas and Shasta counties on the western slopes of the Sierra Nevada and Siskiyou County in the Klamath Mountains.

Tuckermannopsis platyphylla (Tuck.) Hale (pl. 10d)

Thallus brown to olive-brown, loosely adnate to suberect on bark, 3–7 cm broad; lobes 3–6 mm wide, surface wrinkled and papillose, margins darker brown and papillose; lower surface light brown, wrinkled, rhizines sparse or lacking. Apothecia common, 2–5 mm wide, disk brown, rim dark brown and papillose. Cortex and medulla K–, C–, P– (fatty acids).

Habitats: On branches of Douglas Fir, pines, and other coni-fers in North Coastal and Montane Forest from 1500 to 7500 ft elevation.

Range: See fig. 80d.

Umbilicaria Hoffm.

Umbilicaria, often called "rock tripe," is a conspicuous foli-ose-umbilicate lichen attached to rocks by a central cord. The black apothecia are usually deeply fissured in the form of con-centric circles; one species, *U. polyrrhiza,* has radial fissures. Another three species, *U. decussata* Vill. (rare on Mt. Shasta), *U. krascheninnikovii,* and *U. virginis,* have a smooth disk and ridges toward the center of the thallus. All species contain gy-rophoric acid (medulla C+ rose). Since these lichens grow on dry sunny rocks, they are very brittle, breaking into fragments when collected unless first moistened.

The most common species is *U. phaea,* which grows on boulders in open areas in almost every county of the state. The other 10 species reported from California are much rarer, sometimes difficult to identify positively, and collected at higher elevations in the Sierra Nevada and Klamath Moun-tains. Two rare species in the Sierra Nevada can be recognized at sight: *U. hirsuta* with soredia around the thallus margin and *U. deusta* (L.) Baumg. with isidia over the upper surface. An-other small, unrelated umbilicate species, *Peltula euploca,* has a sorediate margin (see Crustose Lichens).

One must be careful not to confuse Umbilicarias with *Dermatocarpon*, which is also umbilicate but has black dots (immersed perithecia) over the upper surface and never produces apothecia.

1. Margins of thallus sorediate.
   [2. Thallus about 1 cm broad; medulla C−. . . . .
   . . . . *Peltula euploca*]
   [2. Thallus 2–6 cm broad; medulla C+ rose. . . . .
   . . . . *U. hirsuta*]
1. Margins lacking soredia, smooth to dissected or lobulate.
   [3. Upper surface with fine isidia. . . . . *U. deusta*]
   3. Upper surface smooth to rugose, lacking isidia.
      4. Lower surface smooth and bare to finely papillose.
         [5. Surface covered with black dots (perithecia);
         medulla C−. . . . . *Dermatocarpon*]
         5. Perithecia lacking (black pycnidia when present
         accompanied by apothecia); medulla C+ rose.
            6. Center of thallus reticulately ridged and
            pruinose.
               7. Lower surface brown; ridges mostly in
               center of thallus. . . . .
               . . . . *U. krascheninnikovii*
               [7. Lower surface black; ridges extending
               to the margins. . . . . *U. decussata*]
            6. Center of thallus uniform, not ridged (pruinose only in *U. vellea*).
               8. Thallus deeply dissected and lobed, subfoliose. . . . . *U. polyphylla*
               8. Thallus entire, prostrate on rock, circular.
                  9. Thallus margin smooth. . . . *U. phaea*
                  9. Thallus margin perforated and lacerated. . . . . *U. hyperborea*
      4. Lower surface partly or completely covered with
      rhizines or plates.
         10. Lower surface light pink. . . . . *U. virginis*
         10. Lower surface dark brown to black.
            11. Margins lacerated. . . . . *U. torrefacta*
            11. Margins smooth to sparsely lobulate, often
            with a projecting mat of black rhizines.

[12. Apothecia with a smooth to concentrically fissured disk. . . . .
. . . . *U. angulata*]
12. Apothecia radially fissured. . . . .
. . . . *U. polyrrhiza*

Umbilicaria hyperborea (Ach.) Hoffm. (fig. 48a)

Thallus blackish brown to brown, loosely adnate, 1–5 cm broad; margins lacerated and perforated, the surface uneven to wrinkled; lower surface black, smooth, lacking rhizines. Apothecia common, about 1 mm wide, disk black with concentric fissures. Cortex K−; medulla C+ red, K−, P− (gyrophoric acid).

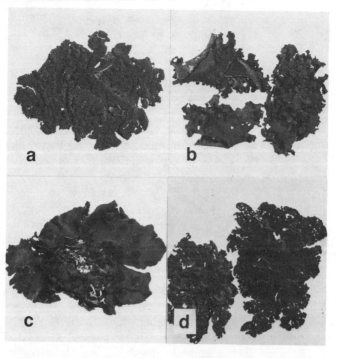

FIG. 48. Species of *Umbilicaria*: (a) *U. hyperborea*; (b) *U. poly-phylla*; (c) *U. polyrrhiza*; (d) *U. torrefacta* (×1).

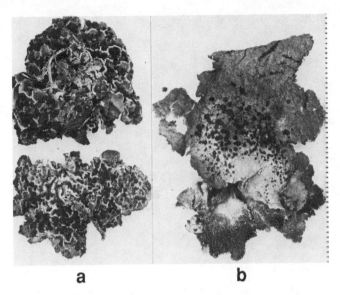

a   b

FIG. 49. Species of *Umbilicaria:* (a) *U. krascheninnikovii;* (b) *U. virginis* (×1).

Habitats: On exposed rocks in Montane and Subalpine Forest from 4000 to 6000 ft elevation.

Range: Rare in Colusa and Tehama counties and from Placer County to Oregon in the Sierra Nevada and Modoc Plateau.

Umbilicaria krascheninnikovii (Sav.) Zahlbr. (fig. 49a)

Thallus dark brownish gray, becoming whitish along ridges at the center, 1–2 cm broad; surface becoming scabrid with age; lower surface smooth gray. Apothecia common, to 0.5 mm wide, disk smooth or with a single ring, black. Medulla K−, C+, KC+ red, P− (gyrophoric acid).

Habitats: On exposed rocks in Montane and Subalpine Forest at 6000 to 8000 ft elevation.

Range: Rare in Santa Barbara, Tehama, and Glenn counties in the North and South Coast Ranges and from Fresno County to

Siskiyou and Modoc counties in the High Sierra and Modoc
Plateau.

Umbilicaria phaea Tuck. (pl. 11a)

Thallus brown, loosely adnate to closely appressed, 2–5 cm
broad; surface smooth; lower surface brown, finely papillate
but lacking rhizines. Apothecia very common, to 1.5 mm
wide, disk black with concentric fissures. Cortex K−; medulla
C+ red, K−, P− (gyrophoric acid).

Habitats: On rock outcrops in Valley and Foothill Woodland
and in Montane and Subalpine Forest from near sea level to
8000 ft elevation.

Range: See fig. 81a.

Umbilicaria polyphylla (L.) Baumg. (fig. 48b)

Thallus dark brown, leathery but easily fragmenting when collected, loosely adnate, 2–3 cm broad; margins laciniate and
dissected, ascending and appearing as lobes, the surface
smooth; lower surface black, smooth, rhizines lacking. Apothecia very rare. Cortex K−; medulla K−, C+ red, P− (gyrophoric acid).

Habitats: On rocks in Montane Forest from 1000 to 3000 ft
elevation.

Range: Rather rare in the Santa Cruz Mountains in the North
Coast Ranges and in the Klamath Mountains.

Umbilicaria polyrrhiza (L.) Fr. (fig. 48c)

Thallus dark brown, leathery, loosely adnate, 2–4 cm broad;
surface smooth; lower surface black, densely rhizinate with
rhizines projecting out along the margins. Apothecia common,
4–5 mm wide, disk black, radially wrinkled. Cortex K−;
medulla K−, C+ red, P− (gyrophoric acid).

Habitats: On rocks in North Coastal and Montane Forest from
2500 to 4500 ft elevation.

PLATE 1

a. *Acarospora chlorophana* (yellow) and *Caloplaca saxicola* (orange).

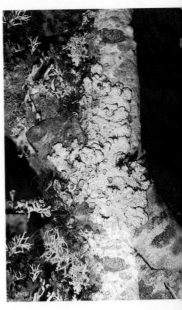

b. *Ahtiana sphaerosporella.*

c. *Bryoria fremontii* (brown) and *Tuckermannopsis canadensis* (yellow).

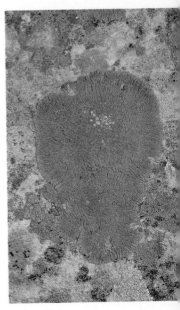

d. *Caloplaca trachyphylla.*

PLATE 2

a. *Candelaria concolor.*

b. *Candelariella rosulans.*

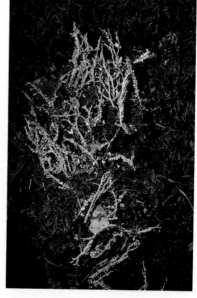

c. *Cladonia fimbriata.*

d. *C. furcata.*

PLATE 3

a. *Cladonia macilenta.*

b. *Collema furfuraceum.*

c. *Dimelaena oreina.*

d. *Evernia prunastri.*

PLATE 4

a. *Flavoparmelia caperata.*

b. *Flavopunctelia flaventior.*

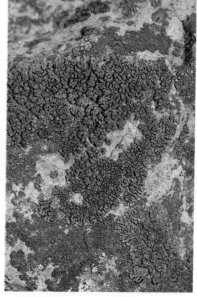

c. *Hypogymnia imshaugii.*

d. *Lecanora mellea.*

PLATE 5

a. *Lecanora muralis.*

b. *Letharia columbiana.*

d. *Melanelia glabra.*

c. *Lobaria pulmonaria.*

PLATE 6

a. *Melanelia subaurifera.*

b. *M. subolivacea.*

d. *Niebla homalea.*

c. *Nephroma resupinatum.*

PLATE 7

a. *Parmelia sulcata.*

b. *Parmelina quercina.*

c. *Parmeliopsis ambigua.*

d. *Parmotrema chinense.*

PLATE 8

a. *Peltigera canina.*

b. *Physconia detersa.*

c. *Platismatia glauca.*

d. *Pseudocyphellaria anomala.*

PLATE 9

a. *Ramalina farinacea.*

b. *R. leptocarpha.*

c. *Rhizocarpon geographicum.*

d. *Rhizoplaca chrysoleuca.*

PLATE 10

a. *Teloschistes chrysophthalmus.*

b. *T. flavicans.*

c. *Tuckermannopsis chlorophylla.*

d. *T. platyphylla.*

PLATE 11

a. *Umbilicaria phaea.*

b. *Usnea rubicunda.*

c. *Xanthoparmelia cumberlandia.*

d. *X. mexicana.*

PLATE 12

a. *Xanthoparmelia taractica.*

b. *Xanthoria elegans.*

c. *X. fallax.*

d. *X. polycarpa.*

Range: Widespread from the Santa Cruz Mountains north-ward to Humboldt County in the North Coast Ranges and Klamath Mountains and Modoc and Shasta counties in the Modoc Plateau.

Umbilicaria angulata Tuck., a rare species in the Sierra Nevada, also has dense black rhizines below but differs in having a concentrically fissured disk.

Umbilicaria torrefacta (Lightf.) Schrad. (fig. 48d)

Thallus variable brown to gray, rather brittle, loosely adnate, 1–6 cm broad, the margins lacerated; surface finely wrinkled; lower surface light to very dark brown, rhizines abundant, flattened. Apothecia common, to 1 mm wide, disk black with concentric rings. Cortex K−; medulla K−, C+ red, P− (gyrophoric acid).

Habitats: On exposed rock outcrops in Montane and Subalpine Forest from 4000 to 7000 ft elevation.

Range: Rare from Inyo County to Modoc and Siskiyou counties in the Sierra Nevada and Modoc Plateau.

Umbilicaria virginis Schaer. (fig. 49b)

Thallus greenish brown, 2–4 cm broad; surface wrinkled and becoming slightly scabrid and pruinose; lower surface pinkish, densely rhizinate, the rhizines light brown. Apothecia very common, 1–1.5 mm wide, disk black, smooth or with one central ring. Medulla K−, C+, KC+ red, P− (gyrophoric acid).

Habitats: On boulders in Subalpine Forest from 6500 to 10,000 ft elevation.

Range: Rare from Fresno County to Alpine County in the High Sierra and in Santa Barbara County in the South Coast Ranges.

Xanthoparmelia (Vain.) Hale

This foliose genus, a segregate of *Parmelia,* includes the narrow-lobed saxicolous species with usnic acid (thallus yellow-green). Cilia are always lacking and the rhizines simple. The apothecia have simple colorless spores. *Xanthoparmelia* is by far the dominant foliose lichen on granites, schists, shale, and other noncalcareous rocks throughout California. Two other common saxicolous yellow-green species, *Dimelaena oreina* and *Lecanora muralis,* have marginal lobes but are actually crustose without a lower cortex and rhizines.

The taxonomy and identification of the genus are not easy for the beginner. While presence of isidia, color of lower surface (brown or black), and degree of adnation distinguish certain species, the most important character is chemistry, which is ideally determined with thin-layer chromatography. Color tests with KOH are helpful but not always definitive at the species level. Without access to TLC equipment you will have to be satisfied with identification to species groups.

Two species, *X. cumberlandia* and *X. mexicana,* are common and collected almost everywhere in the state. The remaining nine species have wide occurrence but are not collected as often. There is no doubt, however, that more species than these will be found in California, especially in the desert regions.

1. Thallus isidiate; apothecia not commonly developed.
    [2. Lower surface uniformly black. . . . . . . *X. conspersa*]
    2. Lower surface uniformly tan to brown.
        [3. Medulla K−. . . . . *X. subramigera*]
        3. Medulla K+ yellow or yellow turning red.
            4. Medulla K+ quickly red (norstictic or salazinic acids); lobes rather broad and rotund at maturity.
                5. Salazinic acid only present. . . . .
                . . . . *X. mexicana*
                [5. Barbatic and salazinic acids present. . . . . .
                . . . . *X. schmidtii*]
            4. Medulla K+ yellow, slowly turning red after several minutes (stictic and norstictic acids); or K+ persistent yellow (psoromic acid); lobes narrower.
            [6. K+ yellow turning red. . . . . *X. plittii*]
            [6. K+ persistent yellow. . . . . *X. kurokawae*]
1. Thallus lacking isidia; apothecia often present.
    7. Medulla K−. *X. novomexicana*

7. Medulla K+ yellow or red.
    [8. Thallus sorediate............... *X. mougeotii*]
    8. Thallus lacking soredia.
        [9. Lower surface uniformly black.............
         .... *X. hypopsila*]
        9. Lower surface tan to brown, sometimes darkening but never uniformly black.
            10. Medulla K+ yellow slowly turning red (stictic and norstictic acids)...........
             .... *X. cumberlandia*
            10. Medulla K+ yellow almost at once turning red (norstictic or salazinic acid).
                11. Thallus loosely attached and easily removed. .... *X. taractica*
                11. Thallus closely attached, difficult to collect free of the rock.
                    12. Salazinic acid present; widespread species. ..... *X. lineola*
                    [12. Norstictic acid present; rare species........... *X. californica*]

Xanthoparmelia cumberlandia (Gyel.) Hale   (pl. 11c)

Thallus yellowish green, closely adnate, 3–10 cm broad; lobes 1–4 mm wide, crowded, older parts often becoming lobulate; lower surface tan, moderately rhizinate. Apothecia common, 2–7 mm wide, disk brown. Cortex K−; medulla K+ yellow turning red, P+ orange (usnic, stictic, norstictic, and constictic acids).

Habitats: On rocks (granite, sandstone, lava) in North Coastal and Montane Forest and in Valley and Foothill Woodland from 500 to 6000 ft elevation.

Range: See fig. 81b.

This common lichen is virtually identical with *X. lineola,* which contains salazinic acid. The color test with K—theoretically K+ more persistent yellow in *X. cumberlandia* and K+ rapidly turning red in *X. lineola*—is not always satisfactory; only a TLC test (or microcrystal test) will positively identify

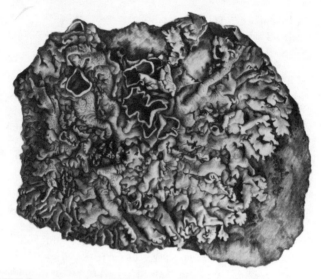

FIG. 50. *Xanthoparmelia lineola* (×2).

the acids. Rare *X. hypopsila* (Müll. Arg.) Hale, known only from Siskiyou County, is externally similar but has a black lower surface.

Xanthoparmelia lineola (Berry) Hale   (fig. 50)

Thallus yellow greenish, closely adnate and not easily removed from rock, 5–8 cm broad; lobes 2–4 mm wide, becoming brown along the edges; lower surface tan, moderately rhizinate. Apothecia very common, 2–5 mm wide, disk brown. Cortex K−; medulla K+ yellow turning red, P+ orange (usnic and salazinic acids).

Habitats: On rock outcrops in Valley and Foothill Woodland from 1000 to 3800 ft elevation.

Range: Widespread but not common from San Diego and Riverside counties north to Sonoma and Butte counties in the North and South Coast Ranges and in the foothills of the Sierra Nevada.

A close relative, *X. californica* Hale, has been collected in

Santa Barbara and Sacramento counties. It contains only norstictic acid.

Xanthoparmelia mexicana (Gyel.) Hale (pl. 11d)

Thallus yellowish green, closely adnate, 3–7 cm broad, fusing
into larger adnate colonies; lobes 2–4 mm wide, isidiate, isidia
dense, coarse, turning brown with age; lower surface light
brown, moderately rhizinate. Apothecia not common, 1–6
mm wide, disk dark brown. Cortex K−; medulla K+ yellow
turning red, P+ orange (usnic and salazinic acids).

Habitats: On rocks and occasionally soil in Valley and Foothill
Woodland and Montane Forest from 500 to 8000 ft elevation.

Range: See fig. 81c.

This species is the commonest representative of the isidiate
Xanthoparmelias. One species, *X. conspersa* (Ach.) Hale, can
be recognized at sight by the black lower surface. The other
species all have a pale brown lower surface and are differentiated primarily by chemistry. A negative color test with K in the
medulla (fumarprotocetraric acid) distinguishes *X. subrami-
gera* (Gyel.) Hale, a widespread species. Rarer *X. plittii* (Gyel.)
Hale reacts K+ yellow slowly turning red (stictic and norstictic
acids), but a microchemical test would be needed to tell it positively from *X. mexicana* (Gyel.) Hale. Very similar *X. kuroka-
wae* (Hale) Hale, a desert lichen, reacts K+ persistent yellow
(psoromic acid). And finally *X. schmidtii* Hale has an unusual
chemistry—barbatic, norstictic, and salazinic acids—and has
been collected only in Tulare County.

Xanthoparmelia novomexicana (Gyel.) Hale

Thallus yellowish green, very closely adnate to adnate, 3–6
cm broad; lobes 1–3 mm wide, crowded; lower surface tan,
moderately rhizinate. Apothecia common, 1–4 mm wide, disk
dark brown. Cortex K−; medulla K−, C−, P+ red (usnic,
fumarprotocetraric, and succinprotocetraric acids).

Habitats: On open rock outcrops in Valley and Foothill Woodland at 800 to 3000 ft elevation.

Range: Fairly common from Tulare County to Mariposa County in the western foothills of the Sierra Nevada.

The K− medullary reaction is sufficient to tell this species from *X. cumberlandia*. Externally it is very similar to *X. lineola*.

Xanthoparmelia taractica (Kremph.) Hale   (pl. 12a)

Thallus yellow-green, loosely adnate, 6–13 cm broad; lobes 2–4 mm wide, linear, dichotomously branched, surface often with black pycnidia; lower surface tan, moderately rhizinate. Apothecia very common, 3–5 mm wide, disk dark brown. Cortex K−; medulla K+ yellow turning red and P+ orange (usnic and salazinic acids).

Habitats: On rocks or soil in Valley and Foothill Woodland, Montane Forest, and Subalpine Forest from 1000 to 8500 ft elevation.

Range: Widespread but not common from San Diego County to Humboldt County in the North and South Coast Ranges and from Kern to Plumas counties inland in the foothills of the Sierra Nevada.

This species is separated from the more common *X. lineola* by difference in adnation. It can usually be removed from rock without damage to the thallus whereas *X. lineola* must be collected with the rock substratum because it is closely adnate. Intermediates, however, cannot always be identified positively.

Xanthoria (Fr.) Th. Fr.

This is a small, narrow-lobed lichen instantly recognized by the deep orange-red color. It is related to crustose *Caloplaca* and fruticose *Teloschistes* since all have two-celled polarilocular spores and contain the orange, K+ purple cortical pigment parietin. It also has a lower cortex and sparse rhizines, characters which differentiate it from the subfoliose Caloplacas (such as *C. saxicola*), which are tightly appressed and lack rhizines.

While Xanthorias can be identified at once in the field, the limits of the species are often poorly defined and will remain so until the genus is revised in North America. For example, *X. candelaria* and *X. fallax,* two sorediate species, intergrade so much in lobe size and configuration that exact identification can be difficult.

The three commonest species, *X. candelaria, X. fallax,* and *X. polycarpa,* are typically found on oak trees. Rare *X. parietina* (L.) Th. Fr., a species more at home in Europe and along the coast of New England, has broad, expanded lobes (1–2 mm wide) and has been collected sporadically from the San Francisco Bay area to Humboldt County. *Xanthoria elegans* and *X. sorediata* are strictly saxicolous species widespread at higher elevations.

1. Thallus lacking soredia; apothecia usually present.
  2. Collected on rocks.
    [3. Lobes closely attached, lacking a lower cortex. . . .
    . . . . *Caloplaca saxicola*]
    3. Lobes adnate with a lower cortex free of the rock.
    . . . . *X. elegans*
  2. Collected on trees and shrubs.
    [4. Lobes barely 0.2 mm wide, separate. . . . .
    . . . . *X. ramulosa*]
    4. Lobes 0.5–2 mm wide, crowded.
      5. Lobes 0.5–1 mm wide, crowded into small, compact thalli less than 1 cm wide with numerous apothecia. . . . . *X. polycarpa*
      [5. Lobes expanded, 1–2 mm wide; thallus 2–4 cm broad; apothecia not crowded. . . . *X. parietina*]
1. Thallus sorediate or sorediate-pustulate; apothecia rare.
    [6. Soredia originating from laminal coarse pustules. . . . .
    . . . . *X. sorediata*]
  6. Soredia powdery, produced along lobe margins.
    7. Lobes 1 mm or more wide, rather coarse; soralia marginal in labriform patches. . . . . *X. fallax*
    7. Lobes more finely divided, less than 1 mm wide; soralia apical or marginal.
      8. Lobes adnate to suberect with a lower cortex.
      . . . . *X. candelaria*
      [8. Lobes closely attached, short, lacking a lower cortex. . . . . *Caloplaca cirrochroa*]

FIG. 51. *Xanthoria candelaria* (×3).

Xanthoria candelaria (L.) Th. Fr. (fig. 51)

Thallus orange to yellowish orange, closely adnate, up to 2 cm broad, the colonies often fusing; lobes 0.2–0.5 mm wide, becoming suberect, apically dissected, soredia scattered but mostly apical; lower surface white, rhizines sparse. Apothecia lacking. Cortex K+ deep purple; medulla K−, C−, P− (parietin).

Habitats: On oaks and other broadleaf trees, rarely on rocks, in North Coastal and Montane Forest and in Valley and Foothill Woodland from near sea level to 6000 ft elevation.

Range: See fig. 81d.

The typical habitat for this species is twigs of oaks. It often occurs with *X. polycarpa* and Physcias.

Xanthoria elegans (Link) Th. Fr. (pl. 12b)

Thallus orange to dark orange, closely adnate, 1–2 cm wide but colonies often fusing to cover large areas of rock; lobes narrow, crowded, 0.5–1 mm wide; lower surface white with a few rhizines, cortex distinctly developed. Apothecia common, 0.5–1.5 mm wide, disk orange red; spores colorless, polarilocular, 6–7 × 10–12 μm. Cortex K+ purple; medulla K−, C−, P− (parietin).

Habitats: On exposed rocks (especially limestone and volcanic rocks) in Montane and Subalpine Forest from near sea level to 12,000 ft elevation.

Range: Rare in Riverside, Santa Barbara, San Bernardino, and Ventura counties in southern California, the Santa Cruz Mountains in the North Coastal Ranges, and from Inyo County to Modoc County in the Sierra Nevada and the Modoc Plateau.

This is a much rarer lichen than *Caloplaca saxicola,* which differs chiefly in lacking a lower cortex. *Xanthoria sorediata* (Vain.) Poelt is also very similar but easily distinguished by the coarse soredia on the upper surface. It has the same range as *X. elegans* but often grows under rock overhangs.

Xanthoria fallax (Hepp) Arn. (pl. 12c)

Thallus deep orange to greenish yellow, closely adnate, 1–2.5 cm broad, fusing into larger colonies; lobes 1 mm or less wide, sorediate along margins and near lobe tips, forming labriform patches; lower surface white with sparse rhizines. Apothecia lacking. Cortex K+ deep purple; medulla K−, C−, P− (parietin).

Habitats: On oaks and other broadleaf trees in Valley and Foothill Woodland from near sea level to 5000 ft elevation.

Range: Rather common in forested areas throughout the state, excepting the North Coastal Forest and High Sierra.

Xanthoria polycarpa (Hoffm.) Rieb. (pl. 12d)

Thallus bright orange, closely adnate, about 1 cm broad; lobes narrow, crowded, about 0.5 mm wide, dissected at the tips; lower surface white, sparsely rhizinate. Apothecia very common, to 1.5 mm wide, disk bright orange. Cortex K+ deep purple; medulla K−, C−, P− (parietin).

Habitats: On twigs of oak and other broadleaf trees in North Coastal and Montane Forest and in Valley and Foothill Woodland from near sea level to 6500 ft elevation.

Range: Common in forested areas throughout the state, except the Modoc Plateau.

This is an extremely variable lichen which may actually represent several species. One segregate, *X. ramulosa* (Tuck.) Herre, is a very rare species on shrubs along the southern California coast; it has very narrow lobes (less than 0.3 mm wide). An apparently undescribed *Caloplaca* species also mimics *X. polycarpa* but lacks a lower cortex; it covers lower branches of roadside oaks in the southern half of the San Joaquin Valley.

2 • FRUTICOSE LICHENS

Fruticose lichens are the most conspicuous lichens in California and occur in the same habitats as foliose lichens. The thallus consists of round or flattened branches. Depending on the genus, the thallus grows free, dangling from tree branches (fig. 7), is attached at the base and either tufted or pendulous, or grows loose or attached on soil or mosses. Two genera, *Letharia* and *Usnea,* have dense cords in the center of the branches, while the upright stalks of *Cladonia* are hollow (fig. 56).

As with foliose lichens, color is a critical character. Common colors are brown, greenish yellow, or whitish gray, rarely orange. Presence or absence of soredia and isidia is also very important for species identification.

Trying to use the keys and assign names to fruticose lichens can be difficult and frustrating. Large, widespread genera such as *Bryoria, Cladonia, Ramalina,* and especially *Usnea* have vague species limits with considerable morphological variation in single populations. Moreover, the basic taxonomy of most of these groups has not been well studied and there is often confusion on the correct species names.

Key to Fruticose Genera

1. Thallus and apothecia (if present) orange; cortex reacting K+ purple.
   2. Thallus tufted on trees. . . . . *Teloschistes*
   2. Thallus forming mats on rock. . . . .
   . . . . *Caloplaca coralloides*

1. Thallus brown, gray, white, or greenish yellow; cortex K−
   or K+ yellowish.
   3. Thallus uniformly brown to blackish brown.
      4. Thallus prostrate to subpendulous on rock, matted.
         .... *Pseudephebe*
      4. Thallus tufted to pendulous or free-growing mostly
         on trees or rocks, not matted.
         5. Branches with a single pseudocyphellate groove
            running their length (visible with hand lens). ...
            .... *Sulcaria*
         5. Branches lacking pseudocyphellae or pseudo-
            cyphellae scattered, round to elliptical.
            6. Thallus rather soft, tufted to pendulous on
               trees or rarely rocks. .... *Bryoria*
            6. Thallus brittle, growing erect on soil. ......
               .... *Coelocaulon*
   3. Thallus gray, white, or greenish yellow to chartreuse.
      7. Thallus greenish yellow to chartreuse.
         8. Thallus (podetia) intricately branched; surface
            dull, fibrous (use hand lens), cortex lacking;
            growing free on soil. .... *Cladina*
         8. Thallus unbranched to moderately branched; sur-
            face shiny, cortex present; attached to trees or
            rocks.
            9. Branches round in cross section.
               10. Central cord present (examine with hand
                  lens). .... *Usnea*
               10. Cord lacking, center of branches loosely
                  filled with hyphae. .... *Alectoria*
            9. Branches irregularly flattened.
               11. Thallus chartreuse to deep yellow. .....
                  .... *Letharia*
               11. Thallus pale green to yellowish green.
                  12. Thallus soft, grooved and white be-
                     low. ....*Evernia*
                  12. Thallus rather stiff and often brittle,
                     lower surface similar to upper.
                     13. Branches thin, often with elon-

gate striations (use hand lens); collected on trees, less commonly on rocks, throughout the state. . . . . *Ramalina*
13. Branches coarse and stout, without striations; usually collected near the ocean. . . . . *Niebla*

7. Thallus greenish gray to white.
   14. Thallus (podetium) hollow (slice open with razor blade and examine with hand lens), very brittle when dry.
       15. Primary squamulose thallus present; podetia simple to sparsely branched or cup-shaped, more or less firmly attached to soil, wood, or humus. . . . . *Cladonia*
       15. Primary squamules lacking, the podetia growing free on soil, usually richly branched.
           16. Surface of podetia greenish, shiny, and corticate, with scattered squamules. . . . . *Cladonia furcata*
           16. Surface of podetia whitish, dull, and fibrous, without a cortex and lacking squamules. . . . . *Cladina*
   14. Thallus solid, usually rather leathery (brittle only in *Sphaerophorus* and *Stereocaulon*).
       17. Apothecia borne on short stalks (0.5 – 1 cm high).
           18. Apothecia black, round to elongate. . . . . *Pilophoron*
           18. Apothecia brown, round to flattened. . . . . *Baeomyces*
       17. Apothecia (if present) produced on surface or tips of branches.
           19. Branches densely covered with tiny squamule-like phyllocladia. . . . . *Stereocaulon*
           19. Branches smooth, lacking phyllocladia.
               20. Branches round in cross section.

21. Thallus yellowish to whitish
mineral or brownish. . . . . . .
. . . . *Sphaerophorus*
21. Thallus white.
  22. Thallus short and stubby,
    1–3 cm high, C+ red. . . .
    . . . . *Schizopelte*
  22. Thallus long with thin
    branches 2–6 cm long,
    C–. . . . .
    . . . *Dendrographa minor*
20. Branches flattened in cross section.
  23. Branches flat at the base but
    rounded at the tips; surface
    C–. . . . *Dendrographa minor*
  23. Branches uniformly flattened;
    surface C+ red or C–.
    24. Marginal soralia present,
      reacting C+ red; apothe-
      cia lacking. . . . . *Roccella*
    24. Soredia lacking; apothecia
      usually numerous.
      25. Surface of thallus
        C–. . . . .
        . . . . . *Dendrographa*
      25. Surface of thallus C+
        red. . . . . . . *Roccella*

Alectoria Ach.

This well-known fruticose genus has recently been subdivided into four different genera (Brodo and Hawksworth, 1977). The brown or blackish species are now classified in *Bryoria*, *Pseudephebe*, and *Sulcaria*, all of which occur in California. *Alectoria* itself is limited to the yellow-green species with brown, one-celled spores. The four species of *Alectoria* in California are tufted to pendulous and white-striate to pseudo-cyphellate. They can be confused only with *Usnea*, which has a distinct cord in the medulla. *Alectoria* is a typical conifer lichen confined to the northern part of the state, but only *A. sarmentosa* is commonly collected.

1. Thallus isidiate, isidiate-spinulate soredia. . . . .
. . . . *A. imshaugii*
1. Thallus smooth, without isidia or spinules.
2. Thallus tufted, short, commonly with apothecia. . . . . .
. . . . *A. lata*
2. Thallus long pendulous, apothecia rare.
3. Medulla loose, C−, KC+ red (alectoronic acid).
. . . . *A. sarmentosa*
[3. Medulla solid, C+ red (olivetoric acid). . . . .
. . . . *A. vancouverensis*]

Alectoria imshaugii Brodo & Hawks.   (fig. 52a)

Thallus yellowish green, tufted, rather stiff, 2−5 cm long; branches 0.4−0.7 mm in diameter, round to slightly flattened and twisted with numerous white-striate pseudocyphellae, isidiate-sorediate. Apothecia rare, up to 3 mm in diameter, disk brown. Cortex K−; medulla K+ yellow orange or K−, C−, P+ yellow or P− (usnic, thamnolic, or squamatic acids and rarely alectoronic acid).

Habitats: On conifers in North Coastal and Montane Forest from near sea level to 6000 ft elevation.

Range: Widespread from Santa Cruz and Glenn counties northward to Del Norte County in the North Coast Ranges and Klamath Mountains.
This species is easily confused with *Usnea* species because of the similar isidioid spinules. However, it lacks a central cord.

Alectoria lata (Tayl.) Linds.   (fig. 52b)

Thallus yellowish green, tufted, stiff, 8−12 cm long, richly branched; branches 0.4−1.0 mm in diameter, round to irregularly flattened, somewhat twisted, with numerous white-striate pseudocyphellae. Apothecia numerous, apical or lateral, disk brown, 3−5 mm in diameter. Cortex K−; medulla K−, C−, KC+ rose, P− (usnic and alectoronic acids).

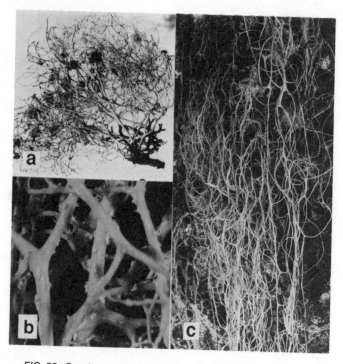

FIG. 52. Species of *Alectoria*: (a) *A. imshaugii*; (b) *A. lata*; (c) *A. sarmentosa* (×1).

Habitats: On conifers in North Coastal and Montane Forest from 2000 to 4000 ft elevation.

Range: Rather rare in Humboldt, Del Norte, and Siskiyou counties in the North Coast Ranges and Klamath Mountains.

Alectoria sarmentosa (Ach.) Ach.   (fig. 52c)

Thallus pale yellowish green, pendulous, flexible, up to 20 cm or more long, little branched; branches 0.2–1 mm in diameter, white-striate to pseudocyphellate, the larger branches twisted. Apothecia rare, disk yellowish, 3–5 mm in diameter. Cortex K−; medulla K−, C−, KC+ rose, P− (usnic and alectoronic acids).

Habitats: On conifers in North Coastal and Montane Forest from sea level to 5000 ft elevation.

Range: Fairly common from the Santa Cruz Mountains northward to Del Norte County in the North Coast Ranges, Cascades, and Klamath Mountains and from Tuolumne County to Shasta County on the western slopes of the Sierra Nevada.

This is by far the commonest species of *Alectoria.* Almost every beginner identifies it as an *Usnea,* but the branches are smooth and lack fibrils and there is no central cord. It may festoon trunks of conifers in the higher mountains but stops at the winter snowpack line. Another pendulous species, *A. vancouverensis* (Gyel.) Brodo & Hawks., has been collected rarely in Humboldt and Del Norte counties; a C+ color test will separate it from *A. sarmentosa.*

Baeomyces Pers.

This inconspicuous lichen is related to *Cladonia* in having a crustose primary thallus which gives rise to short, erect, apothecia-bearing podetia. It differs from *Cladonia* in that the podetia are solid. It is a typical soil consolidator along road banks. Only one species is known from California.

Baeomyces rufus (Huds.) Rabent.   (fig. 53)

Primary thallus crustose-areolate, closely attached to soil, pale greenish gray, 3–5 cm broad; podetia about 2 mm high, solid, simple to divided once, pale flesh-colored, 0.5–1 mm in diam-

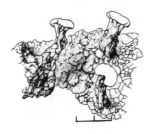

FIG. 53. *Baeomyces rufus* (scale in millimeters).

eter, tipped with light brown apothecia about 1 mm wide. Thallus K+ yellow, P+ orange (stictic acid).

Habitats: On soil in open areas in North Coastal Forest from near sea level to 2000 ft elevation.

Range: Rare but easily overlooked in the Santa Cruz Mountains northward to Humboldt County in the North Coast Ranges.

Bryoria Brodo & Hawks.

Bryoria is a tufted or pendulous brown fruticose lichen with round branches in cross section. It resembles horsehair hanging from tree branches in the conifer forests. The inconspicuous apothecia contain simple colorless spores. It has only very recently been separated from *Alectoria,* which is yellow-green and has brown spores.

Brodo and Hawksworth (1977) recognize 14 species of *Bryoria* in California, many of them rarely collected in the North Coastal Forest and difficult to identify. There is much intergradation of subtle morphological characters and a varied, sometimes involved chemistry requiring very careful TLC tests. We include in the keys what appear to be the commoner species. Students interested in greater detail should consult Brodo and Hawksworth's monograph.

Color tests directly on the finely branched thalli are almost impossible. The best method is to break off some fragments of the branches and place them on white filter paper. A solution of KOH added directly will diffuse out some acids as a deep yellow color. The fragments can also be extracted with a few drops of acetone and P or C applied to the filter paper residue to obtain a spot test.

1. Thallus with powdery, capitate soralia (use hand lens).
    2. Soralia yellow. . . . . *B. fremontii*
    2. Soralia white to greenish.
        [3. Soredia intermixed with isidioid spinules. . . . .
        . . . . *B. furcellata*]
        3. Soredia powdery without spinules. . . . . . *B. glabra*
1. Thallus lacking soredia.

4. Thallus pale grayish brown to tan; extract on filter paper instantly deep yellow with KOH. . . . . *B. capillaris*

4. Thallus greenish or reddish to dark brown; KOH test negative or very pale yellow.

    [5. Collected on rocks at higher elevations. . . . .
. . . . *Pseudephebe minuscula* or *P. pubescens*]

    5. Collected on trees, very rarely on rocks.

        6. Thallus tufted, stiff, 2–3 cm high with terminal apothecia. . . . . *B. abbreviata*

        6. Thallus pendulous, 3–40 cm long, soft; apothecia rarely seen.

        [7. Pseudocyphellae appearing as a long furrow running the length of the branches. . . . . . .
. . . . *Sulcaria badia*]

        7. Pseudocyphellae short and inconspicuous.

            [8. KOH test yellow on filter paper (after acetone extraction). . . . .
. . . . *B. pseudofuscescens*]

            8. KOH test negative.

                9. C test rose on filter paper (after acetone extraction). . . . . . . *B. friabilis*

                9. C test negative.

                    10. Thallus short, 3–8 cm long, brittle with short spinules, matting when wet. . . . . *B. oregana*

                    10. Thallus 6–40 cm long, lacking spinules, soft, not matting when wet.

                        11. Pseudocyphellae sparsely developed, inconspicuous, white. . . . . . *B. fremontii*

                        [11. Pseudocyphellae conspicuously developed, pale yellow (vulpinic acid). . . .
. . . . *B. tortuosa*]

Bryoria abbreviata (Müll. Arg.) Brodo & Hawks. (fig. 54a)

Thallus reddish to dark brown, tufted, stiff, 2–7 cm long; branches to 1 mm in diameter, dull, frequently branched with

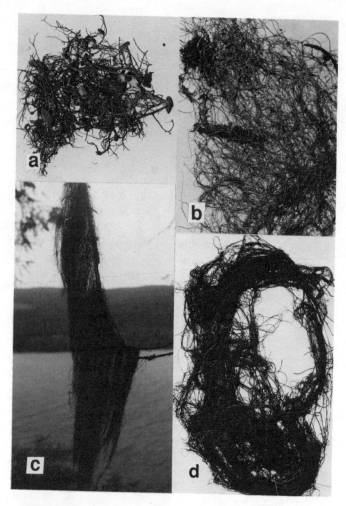

FIG. 54. Species of *Bryoria*: (a) *B. abbreviata*; (b) *B. capillaris*; (c) *B. fremontii*; (d) *B. oregana* (about ×1).

short spinules, becoming somewhat flattened and pitted with age. Apothecia common, terminal, 0.5–2.5 mm wide, disk brown, rim spinulate. Cortex and medulla K−, C−, P− (no substances present).

Habitats: On conifer branches in North Coastal, Montane, and Subalpine Forest from sea level to 6500 ft elevation.

Range: Rather common from San Diego County to Del Norte County in the North and South Coast Ranges and Klamath Mountains and from Kern County to Siskiyou County in the High Sierra and Modoc Plateau.

The stiff, tufted thalli often cover the terminal twigs and branches of spruce and fir, intermixed with *Tuckermannopsis merrillii.*

Bryoria capillaris (Ach.) Brodo & Hawks.   (fig. 54b)

Thallus tannish to pale brown or darkening, pendulous, draping branches loosely, 15–25 cm long; branches smooth, sometimes twisted weakly, 0.1–0.3 mm in diameter; soredia and isidia lacking. Apothecia not seen. Acetone extract on filter paper K+ deep yellow, P+ yellow (alectorialic acid).

Habitats: On conifers in North Coastal and Montane Forest from near sea level to 5000 ft elevation.

Range: Rare from San Luis Obispo County northward to Humboldt and Siskiyou counties in the North Coast Ranges and Klamath Mountains.

This rare species may become so pale that it resembles an *Usnea,* but there is no central cord.

Bryoria fremontii (Tuck.) Brodo & Hawks.   (fig. 54c and pl. 1c)

Thallus dark brown to chestnut, pendulous, 6–40 cm long, often draping tree branches; main branches to 2 mm wide, foveolate and twisted toward the base, ultimate branches less than 1 mm wide, shiny, infrequently branched, rarely sorediate, the soralia yellow (vulpinic acid). Apothecia rare. Cortex and medulla K−, C−, P− (no substances present).

Habitats: On conifers, oaks, and other broadleaf trees in North Coastal, Montane, and Subalpine Forest from near sea level to 8000 ft elevation.

Range: Fairly common and locally abundant from Los Angeles County to Mendocino and Del Norte counties in the North and South Coast Ranges and Klamath Mountains and from Tulare County to Shasta and Modoc counties on the western slopes of the Sierra Nevada, Cascade Mountains, and Modoc Plateau.

This is the most commonly collected *Bryoria* in California. It may hang in festoons from oaks and conifers. The soralia are not always yellow. Another species with vulpinic acid, *B. tortuosa* (Merr.) Brodo & Hawks., has distinct, yellowish pseudocyphellae. Also in the Coast Range, especially in Mendocino and Humboldt counties, one may collect rare *B. pseudofuscescens* (Gyel.) Brodo & Hawks. and *Sulcaria badia,* both of which are also pendulous lichens.

Bryoria friabilis Brodo & Hawks.   (fig. 55a)

Thallus brown to light brown, short pendulous, fine and soft but crumbling in herbarium collections, 6–10 cm long; branches round and smooth, 0.1–0.3 mm in diameter. Apothecia not seen. Cortex K−; medulla K−, C+ rose, P− (gyrophoric acid; use filter paper color test).

Habitats: On trunks of conifers in North Coastal Forest from 1000 to 3000 ft elevation.

Range: Rather rare from the Santa Cruz Mountains northward to Siskiyou County in the North Coast Ranges and Klamath Mountains.

Bryoria glabra (Mot.) Brodo & Hawks.   (fig. 55b)

Thallus dull olivaceous to dark brown, tufted to somewhat pendulous, 6–15 cm long; branches smooth, shiny, 0.1–0.3 mm

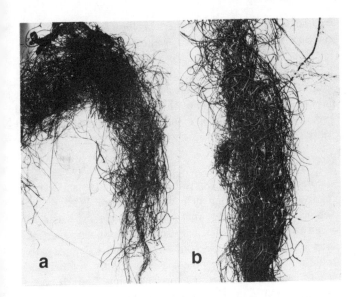

FIG. 55. Species of *Bryoria:* (a) *B. friabilis;* (b) *B. glabra* (about ×1).

in diameter, sorediate, the soralia white, fissural, to 1 mm wide. Apothecia not seen. Cortex K−; soralia K−, C−, P+ red (fumarprotocetraric acid).

Habitats: On conifers in North Coastal Forest from 100 to 2500 ft elevation.

Range: Rare from the Santa Cruz Mountains northward to Del Norte County in the North Coast Ranges.

Related *B. furcellata* (Fr.) Brodo & Hawks., rare in the North Coast Ranges, has isidiate soredia.

Bryoria oregana (Tuck.) Brodo & Hawks.   (fig. 54d)

Thallus dark reddish brown, pendulous, matted when wet but very fragile when dry, 2–10 cm long; branches dull, round to somewhat flattened, 0.3–0.6 mm in diameter, with short spin-ule-like branchlets. Apothecia rare, disk about 1 mm wide. Cortex and medulla K−, C−, P− (no substances present).

Habitats: On branches of conifers in North Coastal, Montane, and Subalpine Forest from sea level to 8000 ft elevation.

Range: Widespread in San Diego and Riverside counties in southern California, becoming more common in Colusa, Glenn, and Tehama counties in the Klamath Mountains and from Tuolumne to Modoc counties in the Sierra Nevada and Modoc Plateau.

Cladina (Nyl.) Harm.

Cladina, popularly known as Reindeer Moss, is a soil-inhabiting fruticose genus often included in *Cladonia*. When well developed it forms a dense mat several feet across. It is recognized by the lack of a primary squamulose thallus and richly branched podetia lacking both a cortex and squamules. Pycnidia and more rarely apothecia occur at tips of branches. While very conspicuous, *Cladina* has only rarely been collected in California. There is only one species.

Cladina portentosa (Duf.) Follm.   (fig. 57a)

Podetia growing free on soil, brittle when dry, pale yellowish green, richly branched, 4–7 cm tall; branches dull, lacking a distinct cortex, the axils closed. Apothecia when present brown, about 0.5 mm in diameter. Thallus K−, P− (perlatolic acid).

Habitats: On sandy soil in North Coastal Forest near sea level.

Range: Locally abundant in Mendocino, Humboldt, and Del Norte counties in the North Coast Ranges.

Cladonia Wigg

Cladonia is one of the first lichens collected by amateurs. It grows on soil, often with mosses, and at the base of trees. It can survive in polluted habitats better than other lichens. The morphology is unique. There is first of all a primary thallus consisting of small, semierect, greenish gray squamules which have a white lower surface. If the plants do not develop beyond these brittle squamules (often the case!), identification to species is often difficult and we call the material simply "*Clado-*

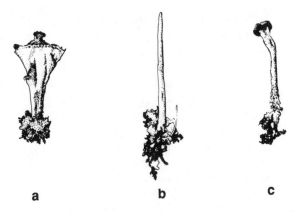

FIG. 56. Types of podetia in *Cladonia:* (*a*) cuplike; (*b*) pointed and sterile; (*c*) pointed with an apothecium at the tip.

nia sp." Usually, however, one can find on the squamules small, erect, branched, pointed, or cup-shaped podetia (fig. 56). These structures are characteristic for each species. Small brown or red apothecia or pycnidia may be found at the tips of the podetia. Presence or absence of soredia and squamules and chemical characters are also used to identify species.

Cladonia is not as common in California as in Oregon and Washington, perhaps because of the smothering effect of the snowpack at higher elevations and the semiarid conditions at lower elevations. Still, 34 species have been reported from the state. The 13 species included here are apparently the commonest ones. Thomson's (1967) study of *Cladonia* in North America may be consulted for more details. Beginners are warned that species identification can be frustrating and difficult because of the great variability in morphology and the apparent "hybridization" between species.

1. Only squamules present.
   2. Lower surface white, lacking a cortex; apothecia lacking . . . . *Cladonia* sp.
   [2. Lower surface gray or darkening, cortex present; adnate apothecia often present on margins or surface of squamules . . . . *Psora* (see Crustose Lichens)]
1. Cup-shaped or pointed podetia arising from the squamules.

3. Podetia solid (cut with razor blade).
    [4. Apothecia elongate, black.......... *Pilophoron*]
    [4. Apothecia round, pale brown. ...... *Baeomyces*]
3. Podetia hollow.
    5. Podetia clearly cup or goblet-shaped, unbranched, with few if any marginal proliferations.
        [6. Surface of cups coarsely areolate; soredia lacking........................ *C. pyxidata*]
        6. Surface of cups with granular or farinose soredia......................... *C. fimbriata*
    5. Podetia blunt or pointed or with weakly developed, shallow cups with well-developed proliferations.
        7. Podetia tipped with red apothecia (or red pycnidia).
            [8. Surface of podetia K−...... *C. bacillaris*]
            8. Surface K+ yellow.
                9. Squamules whitish gray, entire.......
                .... *C. macilenta*
                9. Squamules yellowish, dissected......
                .... *C. transcendens*
        7. Podetia with brown apothecia (or pycnidia) or apothecia lacking.
            10. Shallow, weakly developed cups present.
                11. Cups shallow with central proliferations............... *C. verticillata*
                11. Cups with marginal proliferations or not clearly developed. ....
                .... *C. carassensis*
             10. Podetia lacking cups.
                12. Podetia intricately branched, growing free on soil.
                    [13. Branches lacking a cortex and squamules. .... *Cladina*]
                    13. Branches corticate and with small squamules.
                        14. Soredia lacking. ....
                        .... *C. furcata*
                      14. Soredia present. ....
                        .... *C. scabriuscula*

12. Podetia unbranched or sparsely branched
and crowded.
15. Podetia branched and squamulose.
16. Surface of podetia K−.
17. Branches open, sparsely
squamulose. . . . .
. . . . *C. furcata*
[17. Branches crowded, densely squamulose. . . . .
. . . . *C. crispata*]
16. Surface of podetia K+ yellow.
18. Axils flaring into narrow
cups. . . . *C. carassensis*
[18. Cups not clearly developed. . . . .
. . . . . *C. subsquamosa*]
15. Podetia simple to furcate, squamulose only toward the base; soredia
present.
19. Squamules yellowish, dissected. . . . . . *C. transcendens*
19. Squamules greenish gray, entire. . . . . . . *C. ochrochlora*

Cladonia carassensis Vain.   (fig. 57b)

Primary thallus sparsely developed, the squamules narrow,
1–4 mm long; podetia brownish to whitish gray, 0.6–1 mm in
diameter, 3–4 cm high, brittle, branched several times, sparsely
squamulose, the tips flaring into weakly developed cups with
open axils and rimmed with dark brown apothecia about 1 mm
in diameter. Thallus K+, P+ yellow (thamnolic acid).

Habitats: On soil and over mosses on rocks in North Coastal
Forest and Valley and Foothill Woodland from sea level to
2000 ft elevation.

Range: Widespread from the San Francisco Bay area to Del
Norte County in the North Coast Ranges.
This species is closely related to—and externally barely

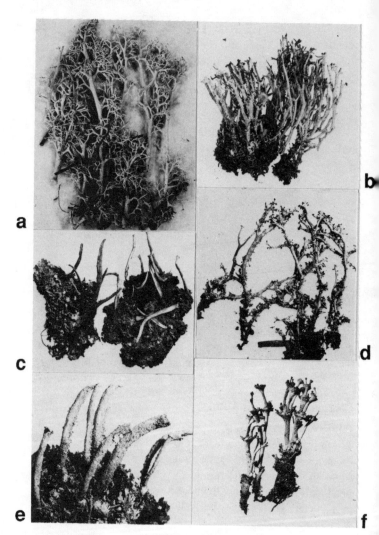

FIG. 57. Species of *Cladina* and *Cladonia*: (*a*) *Cladina portentosa*; (*b*) *Cladonia carassensis*; (*c*) *C. ochrochlora*; (*d*) *C. scabriuscula*; (*e*) *C. transcendens*; (*f*) *C. verticillata* (all near ×1).

distinguished from—*C. crispata* (Ach.) Flot. (K−, P−, squamatic acid present) and *C. subsquamosa* (Nyl.) Vain. (thamnolic acid present but cups barely developed), both of which
occur in the Coast Ranges from the Bay Area to Del Norte
County. They form a difficult group near *C. furcata,* a common species which reacts K− and P+ red (fumarprotocetraric
acid) and has a more richly branched thallus.

Cladonia fimbriata (L.) Fr. (pl. 2c)

Primary thallus sparse to abundant, the squamules 2−5 mm
long, becoming marginally dissected; podetia whitish to greenish gray, cup-shaped but extremely variable, from stout to tall
and narrow; surface in part corticate but always sorediate toward the upper rim, the soredia rather diffuse, powdery. Apothecia rare, brown, about 1 mm in diameter. Thallus surface
K−, P+ red (fumarprotocetraric acid).

Habitats: On soil, on mosses or humus over rocks, and at the
base of trees in North Coastal and Montane Forest and in Valley and Foothill Woodland from sea level to 7000 ft elevation.

Range: Widespread in most of the state outside of deserts.

There are a number of cup Cladonias in California, but their
taxonomy is not yet worked out. For example, *C. major* (Hag.)
Sandst. has large, whitish cups. *Cladonia pyxidata* (L.) Hoffm.
has a well-developed primary thallus and coarse stout cups
covered with areolate granules rather than soredia. *Cladonia
humilis* (With.) Laundon has large corticate areas toward the
base of the podetia. Finally, *C. chlorophaea* (Flk.) Spreng. has
short cups but intergrades to such an extent with the Califor-
nian populations of *C. fimbriata* that they cannot usually be
separated.

Cladonia furcata (Huds.) Schrad. (pl. 2d)

Primary thallus sparsely developed to mostly lacking; podetia
greenish to brownish gray, richly branched and loosely attached to free from the soil, 2−6 cm tall, the surface sparsely
to moderately squamulose, strongly areolate, the axils open.
Apothecia not common, dark brown, to 1 mm in diameter.
Thallus K−, P+ red (fumarprotocetraric acid).

Habitats: On soil or rocks covered with moss and soil in North Coastal and Montane Forest and Valley and Foothill Woodland from 1000 to 4000 ft elevation.

Range: Common from Santa Barbara County northward to Del Norte and Siskiyou counties in the North and South Coast Ranges and Klamath Mountains.

This is one of the few Cladonias which can be identified in the field with confidence because of the free-growing, squamulose thallus.

Cladonia macilenta Hoffm. (pl. 3a)

Primary thallus often poorly developed, the squamules small, to 2 mm wide, the margins dissected, rarely sorediate; podetia simple to sparingly branched, blunt, 3–10 mm high, covered with farinose soredia, often tipped with apothecia. Apothecia bright red, about 1 mm in diameter. Thallus K+, P+ yellow (thamnolic acid).

Habitats: On soil, logs, stumps, and fenceposts in North Coastal Forest, Valley and Foothill Woodland, and Montane Forest from near sea level to 4000 ft elevation.

Range: Rather common from Santa Barbara County to Del Norte County in the North and South Coast Ranges and Klamath Mountains and in Plumas County on the western slopes of the Sierra Nevada.

A morphologically identical species, *C. bacillaris* Nyl., does not react with color tests. It has the same range as *C. macilenta*. When apothecia are lacking, it may still be possible to find small red pycnidia at the tips. Otherwise *C. bacillaris* would be confused with *C. ochrochlora*.

Cladonia ochrochlora Flk. (fig. 57c)

Primary thallus well developed, the squamules short, 1–3 mm wide, the margins becoming dissected and sorediate; podetia simple and pointed, to 2.5 cm tall, covered with farinose to granular soredia. Apothecia lacking. Thallus K−, P+ red (fumarprotocetraric acid).

Habitats: On soil, humus, and rotting logs in North Coastal Forest and Valley and Foothill Woodland into Montane Forest from sea level to 4000 ft elevation.

Range: Very common from Santa Barbara County to Del Norte County in the North and South Coast Ranges and Klamath Mountains and from Plumas County to Shasta County on the western slopes of the Sierra Nevada.

This common species is not always well developed and typical; the podetia are often stunted.

Cladonia scabriuscula (Del.) Nyl. (fig. 57d)

Primary thallus sparsely developed, the squamules 5–8 mm long; podetia greenish white, brittle, dichotomously branched, 2–4 cm tall, moderately to densely covered with squamules, surface of branches becoming decorticated or sorediate. Apothecia rarely developed, brown, to 1 mm in diameter. Thallus K– and P+ red (fumarprotocetraric acid).

Habitats: On mossy banks in North Coastal and Montane Forest from near sea level to 3000 ft elevation.

Range: Common from the San Francisco Bay area northward to Del Norte County in the North Coast Ranges and Klamath Mountains and in Plumas County on the western slopes of the Sierra Nevada.

Cladonia transcendens (Vain.) Vain. (fig. 57e)

Primary thallus well developed, yellowish green, the squamules to 7 mm long, 2–4 mm wide, incised and sorediate at the margins; podetia 1.1–5 cm tall, yellowish, sparingly branched, sorediate. Apothecia rather rare, red, about 1 mm in diameter. Thallus K+ and P+ yellow (usnic and thamnolic acids).

Habitats: On stumps and decaying logs or on mossy banks in North Coastal Forest from near sea level to 2500 ft elevation.

Range: Rather common in Del Norte and Siskiyou counties.

Cladonia verticillata (Hoffm.) Schaer. (fig. 57f)

Primary squamules fairly well developed, greenish brown, linear, 2–4 mm long; podetia pale brownish gray, 2–8 cm high, narrow with shallow flared cups and central proliferations, the rim with small brown pycnidia; surface shiny, smooth, continuous to green-spotted, a few squamules present. Apothecia not seen. Thallus K+ yellow, P+ red (atranorin and fumarprotocetraric acid).

Habitats: On soil and sand dunes in North Coastal Forest near sea level.

Range: Marin and Mendocino counties in the North Coast Ranges (historical record includes Placer County).

Coelocaulon Link

Coelocaulon is a brown fruticose genus common in the Arctic and northern Rocky Mountains but rare and at its southern limit in California. It is likely to be misidentified as a *Bryoria* because of the color and hollow thallus. The main difference between the two, other than habitat (soil-inhabiting in *Coelocaulon* and mostly corticolous in *Bryoria*), is anatomical: *Coelocaulon* has a double cortex, the outer one parenchymatous (cellular), the inner arranged parallel. Only one species is known in California.

Coelocaulon aculeatum (Schreb.) Link (fig. 58a)

Thallus fruticose, forming small tufts on soil 1–2 cm high, dark brown, brittle, 2–4 cm broad, the branches round, hollow, and sparsely but conspicuously pseudocyphellate. Apothecia not seen. Cortex and medulla K−, P− (only fatty acids present).

Habitats: On soil or mossy banks in exposed areas in North Coastal Forest near sea level.

Range: Rare in the San Francisco Bay area and in the North Coast Ranges.

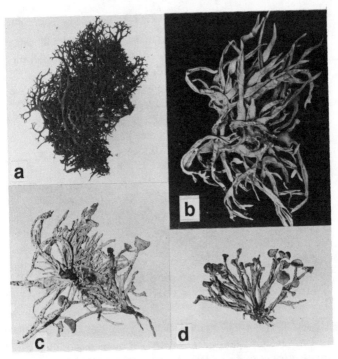

FIG. 58. Species of *Coelocaulon, Dendrographa,* and *Niebla:*
(a) *Coelocaulon aculeatum;* (b) *Dendrographa leucophaea;* (c)
Niebla ceruchis; (d) *N. combeoides* (all near ×1).

Dendrographa Darb.

Dendrographa, a conspicuous fruticose lichen, is related to
Roccella and *Schizopelte,* and like them it grows on shrubs and
rocks near the seashore. It is differentiated by cortical structure
(longitudinally arranged hyphae)—a microscopic character. As
in the other genera the fruiting bodies are ascolocular and the
spores colorless and three-septate. *Dendrographa* is probably
endangered because of habitat disturbance and urbanization.

1. Branches clearly flattened, 0.5–2 mm wide. . . . .
. . . . *D. leucophaea*
[1. Branches mostly rounded, 0.5–1 mm wide. . . . .
. . . . *D. minor*]

Dendrographa leucophaea (**Tuck.**) **Darb.** (fig. 58b)

Thallus whitish gray, pendulous, to 9 cm long, stiff and rather brittle; branches 0.5–2 mm wide, distinctly flattened, the surface smooth to somewhat rugose. Apothecia common, 1–2.5 mm wide, disk convex, whitish gray, pruinose. Cortex and medulla K−, C−, P+ red (fumarprotocetraric acid).

Habitats: On shrubs and rocks at headlands exposed to fog in Coastal Scrub from sea level to 100 ft elevation.

Range: Channel Islands (historical range includes San Diego County northward to Monterey County).

Another species in this genus, *D. minor* Darb., is especially common in the Monterey pine forests at Point Lobos. It also occurs in the Channel Islands, Marin County, and San Francisco County and historically as far south as San Diego. It differs from *D. leucophaea* in having mostly round, more finely divided branches but intergrades with it.

Evernia Ach.

This conspicuous genus is characterized by flattened, light yellowish green, more or less dorsiventral branches, the lower surface being grooved and whitish. The tufted thallus is soft and flaccid in contrast, for example, to the comparatively stiff and brittle thallus of *Ramalina*. There is only one species in California, *E. prunastri,* a very common, easily recognized lichen. Air pollution has largely eliminated it south of Santa Barbara in the San Gabriel and San Bernardino mountains.

Evernia prunastri (**L.**) **Ach.** (pl. 3d)

Thallus pale yellowish green, soft and flaccid, tufted to pendulous, 2–7 cm long; lobes 2–3 mm wide, elongate and often little branched, flattened to canaliculate, surface rugose and pitted, margins with distinct soralia; lower surface white, rugose. Apothecia lacking. Cortex and medulla K−, C−, P− (usnic and evernic acids).

Habitats: On conifers, oaks, and other broadleaf trees, dead wood, and fenceposts in North Coastal Forest, Montane Forest, and Valley and Foothill Woodland from sea level to 5500 ft elevation.

Range: See fig. 73d.

Letharia (Th. Fr.) Zahlbr.

This is the most conspicuous lichen in the Montane Forest of California because of the large bushy, deep lemon-green to chartreuse thallus which covers branches and tree trunks profusely. The branches are irregularly rounded with a wrinkled or rugose surface and internal medullary strands. Apothecia, when present, have a broad brown disk and a lobulate or cornute rim. Spores are colorless and simple. The two species known from the state often occur together and are good indicators of the height of the snowpack on tree trunks, since they do not grow under snow. The common name is Wolf Lichen.

1. Branches sorediate to isidiate-sorediate; apothecia very rare....................................... *L. vulpina*
1. Branches lacking soredia; apothecia almost always present. .... *L. columbiana*

Letharia columbiana (Nutt.) Thoms.   (pl. 5b)

Thallus chartreuse, tufted, 3–11 cm broad, often covering large areas of trunk above the snow line; branches 0.5–3 mm wide, dull, wrinkled and sometimes cracked, with black pycnidia; apothecia always present, 3–17 mm wide, disk brown with a lobulate rim. Cortex and medulla K−, C−, P− (vulpinic acid).

Habitats: On trunk and branches of conifers in North Coastal, Montane, and Subalpine Forest from 3000 to 8000 ft elevation.

Range: Common in forested areas throughout the state, generally above 4000 ft elevation.

Letharia vulpina (L.) Hue   (cover photo)

Thallus golden yellow-green to chartreuse, tufted, 3–15 cm broad, often covering large areas of trunk above the snowpack; branches wrinkled, diffusely sorediate or isidiate-sorediate,

3–8 mm wide. Apothecia very rare, to 20 mm wide, disk brown. Cortex and medulla K−, C−, P− (vulpinic acid).

Habitats: On trunks and dead and living branches of conifers, fenceposts, more rarely on rocks, in North Coastal, Montane, and Subalpine Forest from 1600 to 8500 ft elevation.

Range: See fig. 75a.

Niebla Rundel & Bowler

Niebla is one of the most unusual lichens in the state. This fruticose, mostly saxicolous genus occurs near the ocean or in ocean fog zones, actually reaching maximum development in Baja California. It was long classified in *Ramalina* but differs in having a palisade cortex and leathery, little branched thalli. The spores are colorless and one-septate. The four species in California are most frequently collected from the Channel Islands and northward along the coast to Sonoma County, with *N. cephalota* extending into Humboldt County. Many of the habitats south of Ventura County have been destroyed by urbanization.

1. Apothecia mostly lateral.
    2. Branches strongly flattened with deep transverse cracks.
       . . . . *N. homalea*
    2. Branches irregularly rounded, the surface rugose with white, cottonlike outgrowths with age. . . . *N. ceruchis*
1. Apothecia terminal or lacking.
    [3. Soredia present on branches; branches often black-spotted; growing on trees or rocks. . . . . *N. cephalota*]
    3. Soredia lacking; growing only on rocks.
       4. Apothecia terminal; thallus stout. . . . .
          . . . . *N. combeoides*
       4. Apothecia mostly lateral or lacking; thallus long with linear branches. . . . . *N. homalea*

Niebla ceruchis (Ach.) **Rundel & Bowler**   (fig. 58c)

Thallus greenish yellow, sometimes black-spotted, tufted, 2–4.5 cm long; lobes 0.5–1 mm wide, rounded in cross section, surface cracked and extruding a white cottony substance with age. Apothecia rare, 1–3 mm wide, disk greenish yellow, pruinose. Cortex and medulla K−, C−, P− (usnic and fatty acids).

Habitats: On rocks and twigs along the seashore or in Coastal Scrub near sea level.

Range: Channel Islands and northward along the coast to Sonoma County (historical range includes San Diego, Orange, Los Angeles, and Santa Barbara counties) in the North and South Coast Ranges.

   Niebla cephalota (Tuck.) Rundel & Bowler is a sorediate relative of this species, appearing very much like an *Usnea* but lacking fibrils and a central cord and being softer.

Niebla combeoides (Nyl.) **Rundel & Bowler**   (fig. 58d)

Thallus greenish yellow, tufted, the base blackening, 3–6 cm long; lobes 2–4 mm wide, flattened, to irregularly rounded, surface rugose, rarely extruding a white cottony substance with age. Apothecia common, terminal, to 8 mm wide, disk black and heavily pruinose. Cortex K−; medulla K+ yellow, C−, P+ orange (usnic and stictic acids).

Habitats: On rocks in Coastal Scrub from sea level to 200 ft elevation.

Range: Channel Islands northward to Sonoma County in the North and South Coast Ranges.

Niebla homalea (Ach.) **Rundel & Bowler**   (pl. 6d)

Thallus greenish yellow, tufted, 3–8 cm long; lobes flattened, 1–4 mm wide, surface conspicuously cracked, with marginal black pycnidia. Apothecia abundant, lateral, 1–2 mm wide, disk greenish yellow pruinose. Cortex and medulla K−, C−, P− (usnic and divaricatic acids).

Habitats: On rocks along the seashore in Coastal Scrub up to 600 ft elevation.

Range: Channel Islands north to Sonoma County (historical range includes San Diego and Los Angeles counties) in the North and South Coast Ranges.

Pilophoron (Tuck.) Th. Fr.

This small genus is characterized by a crustose primary thallus which gives rise to stalked fruiting bodies, much as in *Baeomyces*. The apothecia are black and contain simple colorless spores. *Pilophoron* is best developed in the moist coniferous forests of Oregon and Washington, and only one species has been found in California. It grows on rocks in the North Coastal Forest.

Pilophoron aciculare (Ach.) Nyl. (fig. 59)

Podetia erect, firmly attached to rocks, 1–1.5 cm high, little branched, solid, 0.5–1 mm in diameter, the surface finely areolate; primary thallus crustose, greenish gray, 5–10 cm broad. Apothecia black, terminal, elongate, 1–2 mm long; spores simple, colorless. Thallus K+, P+ yellow, C− (atranorin and zeorin).

FIG. 59. *Pilophoron aciculare* (×2).

Habitats: On rocks in moist redwood forest in the North Coastal Forest from near sea level to 1000 ft elevation.

Range: Rare from Sonoma County to Del Norte County in the North Coast Ranges.

Pseudephebe Choisy

Pseudephebe is a small, brown to black fruticose lichen with round to flattened, solid branches. It was formerly classified with *Alectoria*, but the spores are colorless and simple. The two species in California are well-known arctic-alpine lichens, occurring on exposed or wind-swept outcrops at higher elevations, often growing together.

1. Thallus loosely adnate to subpendulous; basal branches usually round; apothecia very rare. . . . . *P. pubescens*
1. Thallus closely attached to appressed; basal branches flattened, apothecia common. . . . . *P. minuscula*

Pseudephebe minuscula (Nyl.) Brodo & Hawks.   (fig. 60a)

Thallus brownish black, closely adnate, 2–6 cm broad; branches shiny, very small, compressed, the lower surface dark brown. Apothecia common, 1–3 mm wide, disk black. Cortex and medulla K−, C−, P− (no substances present).

Habitats: On exposed outcrops in Montane and Subalpine Forest from 6000 to 8000 ft elevation.

Range: Widespread from Fresno County north to Modoc County in the High Sierra and Modoc Plateau.

Pseudephebe pubescens (L.) Choisy   (fig. 60b)

Thallus brownish black, loosely adnate, 1–7 cm broad; branches shiny, rounded, 0.1–0.3 mm in diameter, frequently branched. Apothecia rarely seen. Cortex and medulla K−, C−, P− (no substances present).

Habitats: On exposed rock outcrops in Montane and Subalpine Forest from 4000 to 8000 ft elevation.

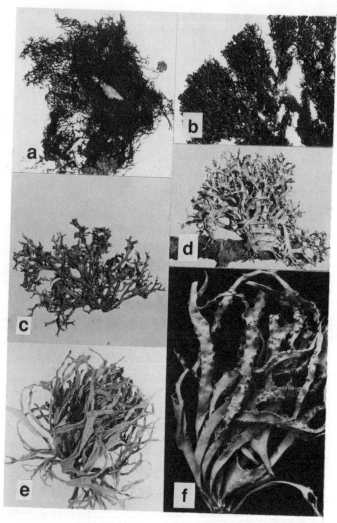

FIG. 60. Species of *Pseudephebe*, *Ramalina*, and *Roccella*: (a) *Pseudephebe minuscula*; (b) *P. pubescens*; (c) *Ramalina dilacerata*; (d) *R. pollinaria*; (e) *R. subleptocarpha*; (f) *Roccella fimbriata* (all near ×1).

Range: Locally common from Fresno County to Siskiyou and Modoc counties in the Sierra Nevada, Cascades, Klamath Mountains, and Modoc Plateau.

Ramalina Ach.

Ramalina is one of the classic fruticose lichen genera. It can be recognized at sight in the field by the tufted or pendulous thallus, flattened branches, and yellow-green color. The distinctive apothecial character is the colorless, two-celled spores. *Ramalina* often occurs with similar-appearing *Usnea,* which has round branches and a central cord, or *Alectoria,* which has round branches and lacks a cord. Very common *Evernia prunastri* seems to be close but has a white, grooved lower surface and a soft thallus.

Many species of *Ramalina* have been reported from California, but the identifications are often doubtful because the genus is notoriously difficult and there are no modern revisions. We are treating the more obvious species here with the realization that some specimens will not match the key well, if at all.

The most spectacular species in the genus, Lace Lichen, *R. menziesii* (*R. reticulata* in older lists), could lay claim to being the unofficial state lichen. It drapes oaks and conifers in festoons from Los Angeles to the Oregon border; it is best developed in fog zones. The most widespread species is *R. farinacea,* a characteristic member of the oak community.

1. Thallus sorediate.
   [2. Thallus soft; lower surface white and grooved. . . . . . .
   . . . . *Evernia prunastri*]
   2. Thallus fairly brittle; lower surface the same color as upper, not grooved.
      [3. Soralia rather diffuse over the surface; branches to 5 mm wide. . . . . *R. duriae*]
      3. Soralia restricted to margins or tips of branches; branches narrow, 1–3 mm wide.
         4. Soralia mostly at tips of branches or produced on the inner surface.
            5. Base of thallus entire. . . . . . . *R. pollinaria*
            [5. Base of thallus perforated. . . . . *R. roesleri*]

4. Soralia linear or orbicular along lobe margins.
 6. Base of thallus yellow-green; thallus rather stiff; medulla K−, P−.................
  .... *R. subleptocarpha*
 6. Base of thallus blackening; thallus rather soft; medulla P+ orange, K+ yellow red, or K−.
  7. Medulla K−.......... *R. farinacea*
  [7. Medulla K+ yellow turning red.....
  .... *R. reagens*]
1. Thallus lacking soredia.
 8. Thallus forming a perforated network, pendulous and draped on branches.................. *R. menziesii*
 8. Thallus tufted to pendulous but attached at the base; branches not perforated as a network.
  9. Apothecia mostly lateral; thallus to 9 cm long, solid and flattened................... *R. leptocarpha*
  9. Apothecia terminal; thallus 1–2 cm long, hollow and perforated................... *R. dilacerata*

Ramalina dilacerata (Hoffm.) Hoffm.

(fig. 60c)

Thallus pale yellow-green, fragile, tufted, 1–2 cm long; branches round to flattened, appearing inflated, hollow, about 1 mm in diameter, more or less regularly perforated with elongate holes. Apothecia terminal, 3–5 mm wide, disk yellowish tan. Cortex and medulla K−, C−, P− (usnic and divaricatic acids).

Habitats: On broadleaf trees (especially apple trees) in North Coastal Forest from 500 to 2000 ft elevation.

Range: Widespread in Santa Cruz, Mendocino, and Humboldt counties in the Coast Ranges.

Ramalina farinacea (L.) Ach.

(pl. 9a)

Thallus pale yellow-green, tufted to pendulous, 2–8 cm long; branches 1–2 mm wide, flattened, becoming divided at the tips; surface foveolate to longitudinally pitted with marginal orbicular soralia; base darkening at point of attachment. Apothecia very rare. Medulla K−, C−, P+ red (usnic and protocetraric acids).

Habitats: On oaks and other broadleaf trees and on conifers in North Coastal and Montane Forest and Valley and Foothill Woodland from 500 to 3000 ft elevation.

Range: See fig. 80a.

Some specimens of this common lichen react K+ yellow in the medulla. These are classified as *R. reagens* (B. de Lesd.) Culb. (norstictic acid present). Another sorediate species, *R. duriae* (De Not.) Bagl., has wide branches and large marginal and laminal soralia. It occurs rarely from San Diego to Santa Barbara County but may be nearing extinction because of development.

Ramalina leptocarpha Tuck.   (pl. 9b)

Thallus pale yellow-green, tufted to pendulous, 3–9 cm long; branches 0.5 to 3 mm wide, flattened, rugose pitted with some longitudinal striations; apothecia very common, mostly lateral, 1–6 mm wide, disk tan. Cortex and medulla K−, C−, P− (usnic acid).

Habitats: On oaks and other broadleaf trees, rarely conifers, in Valley and Foothill Woodland and North Coastal Forest from sea level to over 1000 ft elevation.

Range: Common from San Diego County to Humboldt County in the North and South Coast Ranges.

Ramalina menziesii Tayl.   (fig. 7)

Thallus yellow-green, pendulous and often covering entire trees, 9 to 35 cm long or more; branches 1–10 cm wide, variable, simple to expanded and perforate and netlike, the surface with white striations. Apothecia common, 1–3 mm wide, scattered on surface of wider lobes, disk pale yellow-brown. Cortex and medulla K−, C−, P− (usnic acid).

Habitats: On oaks and other broadleaf trees, shrubs, and conifers in Valley and Foothill Woodland and North Coastal Forest from near sea level to 3500 ft elevation.

Range: See fig. 80b.

Popularly known as Lace Lichen, this is the most conspicuous lichen in Valley and Foothill Woodland. The best development of this Spanish Moss–like species, however, is on hillsides facing the Pacific Ocean. Drier, valley-facing oak forests lack it entirely. A form in fog zones, as can be seen at Patrick's Point State Park, becomes brownish and has narrow branches with few obvious reticulate perforations. It may not be recognized at first as a *Ramalina*.

Ramalina pollinaria (Westr.) Ach.   (fig. 60d)

Thallus pale yellowish green, tufted, rather fragile, 2–3 cm long; branches flattened, shiny, 1–2 mm wide, smooth, the tips bursting open and becoming sorediate on the inner surface. Apothecia not seen. Cortex and medulla K−, C−, P− (usnic and evernic acids).

Habitats: On twigs of conifers and broadleaf trees in North Coastal Forest near sea level.

Range: Santa Barbara County to Humboldt County in the Coast Ranges.

Ramalina subleptocarpha Rundel & Bowler   (fig. 60e)

Thallus yellowish green tufted to pendulous, 3–6 cm long; branches 1–3 mm wide, shiny, flattened, slightly pitted, sorediate, soralia round to elongate, mostly marginal. Apothecia not seen. Cortex and medulla K−, C−, P− (usnic acid).

Habitats: On oaks and other broadleaf trees and fenceposts in North Coastal Forest and Valley and Foothill Woodland from 1000 to 2500 ft elevation.

Range: Rather common from San Diego County to Humboldt County in the North and South Coast Ranges.

This species is considered to be the sorediate form of *R. leptocarpha*. Both species are common on planted trees near the beaches.

Roccella DC.

Roccella is a small but unusual genus of fruticose lichens which grow in Mediterranean climates, usually near the ocean, and which were an important source of purple dyes in the dyeing industries in the nineteenth century. The branches are flattened and rather soft. The apothecia differ from other fruticose genera (except closely related *Dendrographa* and *Schizopelte*) in having the asci scattered in locules in the fruiting layer. Spores are colorless and three-septate. The taxonomy of the genus is confused at the present time, and while Tucker and Jordan's catalog (1979) lists seven species we include only two that are common in the Channel Islands southward into Baja California.

1. Thallus sorediate. *R. babingtonii*
1. Thallus lacking soredia. . . . . *R. fimbriata*

Roccella babingtonii Mont.   (fig. 61)

Thallus white to light gray, tufted to pendulous, 7–12 cm long; branches flattened, 1–4 mm wide with occasional smaller branches toward the base, the surface wrinkled to pitted with abundant white capitate soralia. Apothecia usually not seen. Cortex and soredia K−, C+ red; medulla K−, C−, P− (lecanoric acid).

FIG. 61. *Roccella babingtonii* (×1).

Habitats: On trees, shrubs, or rocks in Coastal Scrub near sea level.

Range: Channel Islands and San Diego County in the South Coast Ranges.

Roccella fimbriata Darb.
(fig. 60f)

Thallus grayish white, suberect to pendulous, 7–14 cm long, rather soft; branches flattened, 1–4 (rarely 8) mm wide, smooth to rugose with age. Apothecia very common, 1–2.5 mm wide, disk black, heavily pruinose. Cortex K−, C+ red; medulla K−, C−, P− (lecanoric acid).

Habitats: On rocks, especially volcanic headlands exposed to fog, in Coastal Scrub from near sea level to 200 ft elevation.

Range: Channel Islands (historic range includes San Diego County).

Schizopelte Th. Fr.

This small, grayish white fruticose lichen has soft, rounded, solid branches. It is related to *Dendrographa* and *Roccella* in having ascolocular apothecia (pseudoapothecia) with three-septate brown spores. Highly restricted to the Channel Islands, it covers the steep cliffs at the boat landing at Anacapa Island.

Schizopelte californica Th. Fr.
(fig. 62a)

Thallus grayish white, growing on shrubs or firmly attached to rock, soft and leathery but breaking apart when preserved, 1–3 cm high; branches round to irregular in cross section, solid, little branched. Apothecia terminal, 5–10 mm wide, disk black to white pruinose. Cortex C+ red; medulla C− (erythrin, lecanoric acid, schizopeltic acid).

Habitats: On shrubs, especially Box-thorn (*Lycium*), and on cliffs in headlands in fog zones in Coastal Scrub near the coast.

Range: Common in the Channel Islands, rarer in Los Angeles and Ventura counties.

FIG. 62. Species of *Schizopelte, Sphaerophorus, Stereocaulon, Sulcaria,* and *Usnea:* (a) *Schizopelte californica;* (b) *Sphaerophorus globosus;* (c) *Stereocaulon intermedium;* (d) *Sulcaria badia;* (e) *Usnea arizonica;* (f) *U. cavernosa* (all near ×1).

Sphaerophorus Pers.

Sphaerophorus has a yellowish greenish gray to tan or brown thallus with round, solid branches. At first it resembles a *Cladonia,* which has hollow branches. The apothecia are unusual since the fruiting layer breaks down and the brown spores form a blackish mass within the disk, a mazaedium identical with that of *Coniocybe* and *Cyphelium.* The only species in California, *S. globosus,* grows in tufts on trunks of redwood and other conifers and is highly restricted to the humid, foggy North Coastal Forest.

Sphaerophorus globosus (Huds.) Vain.   (fig. 62b)

Thallus greenish tan or yellowish to brownish gray, erect to sub-pendulous, 3–7 cm long, stiff; branches round, 0.5–1 mm in diameter, smooth and shiny with numerous tiny side branches. Apothecia at tips of branches, infrequent, 2 mm wide, disk black, convex, the rim enclosing the disk. Cortex and medulla K−, C−, P+ yellow (sphaerophorin, squamatic acid, and thamnolic acid).

Habitats: On conifers, stumps, or dead branches in the North Coastal Forest from sea level to 3000 ft elevation.

Range: Rather common from the Santa Cruz Mountains northward into Oregon in the North Coast Ranges.

Stereocaulon Hoffm.

This fruticose genus is distinguished by the solid branches (pseudopodetia) covered with tiny, squamule-like phyllocladia and wartlike cephalodia. The apothecia are terminal or lateral on the branches with colorless three-septate spores. *Cladonia,* which has a superficial resemblance, has hollow branches (podetia) and true squamules. *Stereocaulon* is at the southern limit of its range in the Pacific Northwest and is much better developed in Oregon and Washington. There are not many collections, most coming from the northern half of the state with a few from the Santa Cruz Mountains. Species identification is not easy; it is based mainly on the shape of the phyllocladia and cephalodia and on chemistry.

The commonest species appears to be *S. intermedium*, which grows on rocks in large sheltered outcrops. At least three other species are reported from the state (Tucker, 1973): *S. myriocarpum* Th. Fr. (Sonoma County), *S. sterile* (Sav.) Lamb (San Mateo County), and *S. tomentosum* Fr. (Modoc County). Further information on these species should be sought in Lamb (1977, 1978).

1. Apothecia mostly terminal or lacking; tomentum weakly developed; lobaric acid present (P−).
  2. Pseudopodetia erect; apothecia usually present along with conspicuous cephalodia. . . . . . . *S. intermedium*
  [2. Pseudopodetia prostrate; apothecia lacking and cephalodia scarce and inconspicuous. . . . . *S. sterile*]
1. Apothecia lateral and terminal; tomentum sparse to well developed; stictic acid present (K+, P+ yellow).
  [3. Phyllocladia in part coralloid; tomentum poorly developed. . . . . *S. myriocarpum*]
  [3. Phyllocladia verrucose to squamulose; tomentum very dense. . . . . *S. tomentosum*]

Stereocaulon intermedium (Sav.) Magn.   (fig. 62c)

Pseudopodetia whitish gray, saxicolous, 2–5 cm high; branches woody, 1–2 mm wide, bare or with thin tomentum, covered with numerous branched phyllocladia; cephalodia large and conspicuous, pale greenish or brownish gray, the surface verrucose. Apothecia mostly terminal, dark brown, 1–2 mm wide. Surface of pseudopodetia K+ yellow, C−, P+ pale yellow (atranorin and lobaric acid).

Habitats: On rocks in large sheltered outcrops in North Coastal and Montane Forest from near sea level to 3000 ft elevation.

Range: Rare from Marin County northward to Del Norte County in the North Coast Ranges and Klamath Mountains.

Sulcaria Bystr.

Sulcaria is a brown fruticose lichen closely allied to *Bryoria* and *Alectoria*. It is distinguished by long, furrowed pseudocyphellae which run the length of the branches, one–three-septate spores, and presence of atranorin. There is one species in California, which might be misidentified as *Bryoria fremontii* without careful study.

Sulcaria badia Brodo & Hawks. (fig. 62d)

Thallus orange to chestnut brown with paler brown main branches, pendulous, rather soft, 10–30 cm long; branches shiny, twisted with age, 0.5–1 mm wide, with long furrowed pseudocyphellae appearing as a vertical groove. Apothecia lacking. Cortex K+ yellow; medulla K−, C−, P− (atranorin).

Habitats: On oaks and conifers in North Coastal Forest from 1000 to 2000 ft elevation.

Range: Rare in Santa Cruz County to Mendocino County in the North Coast Ranges.

Teloschistes Norm.

This orange fruticose genus is one of the most conspicuous lichens in California, although it has become quite rare in the past century and may well face extinction (as has *T. villosus* (Ach.) Norm. already) because of urbanization. The branches are thin, round to flattened, often spinulate, with terminal apothecia when fertile. The spores are polarilocular.

1. Thallus sorediate; apothecia lacking. . . . . *T. flavicans*
1. Thallus lacking soredia but almost always with apothecia.
    2. Thallus closely tufted, 1–2 cm high. . . . .
    . . . . *T. chrysophthalmus*
    [2. Thallus loosely tufted to suberect, 3–6 cm high. . . . .
    . . . . *T. exilis*]

Teloschistes chrysophthalmus (L.) Th. Fr. (pl. 10a)

Thallus pale orange to gray, tufted, 1.5–2.5 cm high, the branches about 1 mm wide, flattened with a wrinkled surface; lower surface lighter in color than the upper. Apothecia very common, mostly terminal, 2–5 mm wide, disk orange, the rim often spinulate. Cortex K+ purple; medulla K−, C−, P− (parietin).

Habitats: On oak in Valley and Foothill Woodland and Coastal Scrub from 500 to 2000 ft elevation.

Range: Rare from Santa Barbara County to the San Francisco Bay area in the South Coast Ranges (historical range extends south to Orange County) and on the Channel Islands.

Teloschistes flavicans (Sw.) Norm.   (pl. 10b)

Thallus yellow-orange to orange, loosely adnate to tufted, 2–4 cm broad; branches less than 0.5 mm wide, rounded, the tips spinulate; soralia orbicular, yellow, rather diffuse. Apothecia lacking. Cortex K+ purple; medulla K−, C−, P− (parietin).

Habitats: On conifers and other trees in Coastal Scrub near sea level.

Range: Rare in Santa Barbara County and the Channel Islands north to Marin County in the South Coast Ranges (historical range includes San Diego and Sonoma counties).

    A similar but smaller nonsorediate species, *T. exilis* (Michx.) Vain., has been collected in the Santa Cruz Mountains and the Channel Islands but is now very rare.

Usnea Adans.

Usnea is a widespread, easily recognized fruticose genus, often called old man's beard. It is characterized by the branching, tufted to pendulous thallus and a dense, threadlike cord in the medulla (see fig. 5). Many species produce fibrils—short thin branches which seem to intergrade with isidia. Papillae, small bumps visible with a hand lens (see fig. 63a), are also commonly produced on the branches. These papillae may give rise to soredia or patches of mixed soredia-isidia. Other species have smooth branches without fibrils or papillae. When found in fruit, the apothecia contain simple, colorless spores.

    The identification of *Usnea* species is often extremely difficult or even impossible. The intergradation of vegetative characters is troublesome, and the chemistry is varied with each species having several different populations. Moreover, many of the species names in herbarium collections are incorrect and there are no satisfactory keys. Tucker and Jordan (1979) list about 30 species for California. We discuss only five here and realize that many specimens will simply have to be left as "*Usnea* sp."

1. Thallus pendulous, hanging loosely from branches and without clear basal attachment.
    [2. Center of branches filled with loose medulla (cut open with razor blade). . . . . *Alectoria sarmentosa*]
    2. Center of branches with a threadlike cord.
        3. Main branches dull, whitish, lacking a cortex. . . . .
            . . . . *U. longissima*
        3. Main branches yellow-green, corticate.
            4. Branches smooth (sometimes foveolate at the base), lacking papillae and soredia. . . . .
                . . . . *U. cavernosa*
            4. Branches papillate, becoming sorediate. . . . . . .
                . . . . *U. ceratina*
1. Thallus tufted to pendulous, firmly attached at the base.
    5. Apothecia terminal, numerous; soredia lacking. . . . . . .
        . . . . *U. arizonica*
    5. Apothecia lacking or lateral and inconspicuous; soredia often present.
        6. Thallus with a reddish cast. . . . . *U. rubicunda*
        6. Thallus yellowish green.
            7. Cord reddish; thallus becoming pendulous. . . . .
                . . . . *U. ceratina*
            7. Cord white; thallus tufted. . . . . *U.* spp.

Usnea arizonica Mot.   (fig. 62e)

Thallus tufted, suberect to pendulous, yellowish green, 6–10 cm long, the branches fibrillose and papillate but lacking isidia. Apothecia common, terminal, 5–10 mm in diameter, disk flesh-colored to white pruinose, the rim densely fibrillose; spores simple, colorless, $5 \times 10 \ \mu m$. Cortex K−; medulla K+ yellow turning red, C−, P+ orange (usnic and salazinic acids).

Habitats: On oaks and other broadleaf trees and on conifers in North Coastal Forest and in Valley and Foothill Woodland into Montane Forest from 1000 to 3000 ft elevation.

Range: Rather common from the San Francisco Bay area to Humboldt County in the North Coast Ranges.

Usnea cavernosa Tuck. (fig. 62f)

Thallus pendulous, draping tree branches, rather soft, pale greenish yellow, 20–60 cm long without clear basal attachment, the branches smooth to foveolate, lacking fibrils, articulately cracked with age. Apothecia lateral, inconspicuous, about 5 mm in diameter, disk yellowish to white pruinose with a short fibrillose rim; spores simple, colorless, 5 × 10 μm. Cortex K−; medulla K+ yellow turning red, C−, P+ orange (usnic and salazinic acids).

Habitats: On oaks and other broadleaf trees and on conifers in open areas in the North Coastal Forest, Valley and Foothill Woodland, and Montane Forest from 1000 to 6000 ft elevation.

Range: Common from the Bay Area northward to Trinity and Humboldt counties in the North Coast Ranges and Klamath Mountains.

Usnea ceratina Ach. (fig. 63a)

Thallus tufted and elongate to pendulous, sometimes lacking distinct basal attachment, yellowish green, 10–50 cm long, the branches papillose, becoming white-spotted and sorediate,

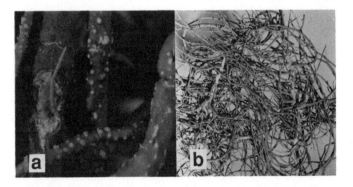

FIG. 63. Species of *Usnea:* (a) *U. ceratina* (×10); (b) *U. longissima* (×1).

sparsely fibrillose, becoming annulate-cracked with age, the cord sometimes turning rusty red. Apothecia not seen. Cortex K−, C−, P− (usnic and diffractaic acids).

Habitats: On oaks and other broadleaf trees in North Coastal Forest and Valley and Foothill Woodland from near sea level to 3000 ft elevation.

Range: Rather common from the San Francisco Bay area northward to Del Norte County in the North Coast Ranges.

Usnea longissima Ach. (fig. 63b)

Thallus pendulous without basal attachment, 10–30 cm long, pale yellowish green, the main branch whitish dull, lacking a cortex, the short fibrillose side branches corticate. Apothecia lacking. Cortex K−; medulla K−, C−, P− (usnic and barbatic acids).

Habitats: On conifers in North Coastal Forest from near sea level to 1000 ft elevation.

Range: Rare from the San Francisco Bay area northward to Humboldt County in the North Coast Ranges.

Usnea rubicunda Stirt. (pl. 11b)

Thallus tufted, erect, yellowish to dull reddish green, 8–12 cm long, the branches finely papillose, white-spotted with rich development of soredia or mixed soredia-isidia. Apothecia not seen. Cortex K−; medulla K+ yellow, C−, P+ pale orange (usnic and stictic acids).

Habitats: On oaks and other broadleaf trees and on conifers in Coastal Scrub, North Coastal Forest, and Valley and Foothill Woodland from near sea level to 3000 ft elevation.

Range: Very common from the Channel Islands and Santa Barbara County northward to Del Norte County in the North and South Coast Ranges.

This common species is typical of the small, tufted, sorediate or sorediate-isidiate Usneas. It is distinguished by the reddish cortical pigment and the presence of stictic acid. Other species in the group lack the pigment, however, and contain salazinic acid, protocetraric acid, barbatic acid, or various unknown substances. These have been called *U. fulvoreagens* (Räs.) Räs., *U. glabrata* (Ach.) Vain., *U. scabiosa* Mot., *U. subfloridana* Stirt., and others, but these names require more study.

3 • CRUSTOSE LICHENS

Crustose lichens have a very simple growth form compared with foliose and fruticose types. The thallus is tightly attached to bark or rock and often penetrates into the substrate, usually making it impossible to collect a crustose lichen free from the substrate. In fact, thallus characters in general play only a small role in the taxonomy of crustose lichens. For example, isidia and soredia are extremely rare.

Crustose lichens, with a few exceptions, must be examined in the laboratory for positive identification. The first step is to determine the type of fruiting body. There are basically two types: open disk-shaped apothecia (fig. 13a) or immersed, black, flask-shaped perithecia (fig. 13b). The former type is by far the more common.

If apothecia are present, examine the disk with a hand lens to see if there is a thalline rim present. This rim will have the same color as the thallus and contain green algae. A proper rim is the same color as the disk and lacks any algae. Next slice off a thin vertical section of the fruiting body with a razor blade. This will take some practice. Mount the section in a drop of water on a microscope slide, cover with a coverslip, and examine under a compound microscope at 100× or higher. The most important microscopic character is the spores (fig. 64), located in sacs (asci) in the fruiting layer (hymenium). Note their size, color (colorless or brown), and septation (number of cross walls if any).

The keys given here will enable a beginner to identify a specimen to genus in most cases. Keys to all of the genera and

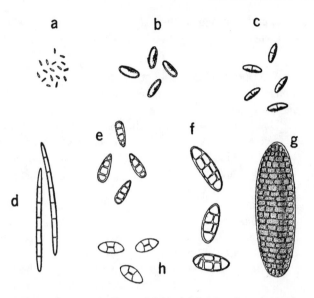

FIG. 64. Spores of lichens: (*a–b*) simple; (*c–e*) transversely septate; (*f–g*) muriform; (*h*) polarilocular.

species listed in the Tucker and Jordan (1979) catalog cannot be presented at this time since our knowledge of crustose lichens in California is very imperfect. If you are interested in pursuing crustose lichens further, additional keys, which include many California species, can be found in Weber (1963) and Wetmore (1967) as well as in the older floras by Hasse (1913) and Herre (1910).

Key to Crustose Lichens

1. Thallus squamulose, consisting of separate but often crowded squamules 1–10 mm long. . . Group A: Crustose-Squamulose Lichens
1. Thallus crustose, lacking any development of squamules.
    2. Thallus with distinct marginal lobes (fig. 18), crustose or chinky-crustose only at the center. . . . . Group B: Crustose, Chinky-Lobate Lichens
    2. Thallus without marginal lobes, continuous and smooth

or uniformly chinky throughout. . . . Group C: Crustose
Lichens

Group A: Crustose-Squamulose Lichens

1. Squamules thin and delicate, up to 1 mm across, with sorediate margins.
    2. Thallus greenish or brownish, reacting C+ red. . . . . . .
    . . . . *Hypocenomyce scalaris*
    2. Thallus whitish gray, C−. . . . . *Normandina pulchella*
1. Squamules thicker and coarser, 3–10 mm long, not sorediate (except in *Cladonia*).
    3. Squamules corticate below; apothecia when present
    marginal. . . . . *Psora*
    3. Squamules lacking a lower cortex; apothecia if present
    mostly laminal (marginal only in *Psorula*).
        4. Squamules greenish gray with a white lower surface.
        . . . . *Cladonia*
        4. Squamules reddish brown to pink or white or brown
        and blackening; lower surface dark.
            5. Squamules with black dots (perithecia) on the
            surface. . . . . *Catapyrenium*
            5. Perithecia absent; apothecia present.
                6. Apothecia sunken; algae dark blue-green. . .
                . . . . *Heppia* and *Peltula*
                6. Apothecia adnate; algae green. . . . *Lecidoma*
                demissum (see *Psora*)

Group B: Crustose, Chinky-Lobate Lichens

1. Thallus and/or apothecial disk orange, reacting K+ purple.
. . . . *Caloplaca*
1. Thallus grayish, brownish, or greenish white or yellow, reacting K− or K+ yellow.
    2. Thallus grayish, brownish, or greenish white, at least at
    the margins.
        3. Thallus with a tan center covered with wartlike
        cephalodia; orbicular soralia common; apothecia
        rare or lacking. . . . . *Placopsis gelida*
        3. Thallus with uniform color, lacking soredia and

cephalodia; apothecia common. . . . .
. . . . *Dimelaena radiata*
2. Thallus yellow to yellowish green.
    4. Apothecial disk dark brown; spores two-celled, brown. *Dimelaena oreina*
    4. Apothecial disk yellowish to pale brown; spores simple, colorless.
        5. Apothecia adnate with a broad yellowish disk and raised thalline rim; eight spores per ascus.
        . . . . *Lecanora muralis*
        5. Apothecia flush with surface of areoles, lacking a rim, or not immediately recognizable; numerous spores per ascus. . . . . *Acarospora chlorophana*

Group C: Crustose Lichens

1. Thallus and/or apothecial disk orange or yellowish orange or rusty red to red, reacting K+ purple.
    2. Disk orange or yellowish orange; spores two-celled, polarilocular. . . . . *Caloplaca*
    2. Disk red; five to seven spores transversely septate. . . . .
    . . . . *Haematomma*
1. Thallus and/or fruiting bodies (apothecia, perithecia, lirellae, or mazaedia) gray to black, white, yellow, or brown, K− or K+ slowly yellowish.
    3. Thallus sorediate, the soralia orbicular or diffuse; fruiting bodies lacking or buried under soralia.
        4. Thallus consisting of a white, pale greenish, gray, or yellow powder.
            5. Powdery granules without a cortex, white, gray, or citrine yellow, on rocks, soil, or bark. . . . . . .
            . . . . *Lepraria*
            5. Granules corticate, scattered on bark, yellow to greenish yellow. . . . . *Candelaria concolor*
        4. Thallus forming a continuous crust; soredia, if present, confined to discrete orbicular soralia about 1 mm in diameter.
            6. Apothecia and spores lacking under soralia.
                7. Thallus C+ red. . . . .
                . . . . *Dirina* (see under *Roccellina*)

   7. Thallus C−. . . . . *Pertusaria*
  6. Apothecia and spores present under soralia.
    8. Spores simple, one or two per ascus. . . . . .
    . . . . *Pertusaria*
    8. Spores densely muriform, one per ascus. . . .
    . . . . *Phlyctis*
3. Thallus lacking soredia (except for *Coniocybe*). (For the following key it is necessary to make a vertical section of the fruiting bodies for examination under a compound microscope.)
  9. Hymenium breaking down as a powdery dark mass of spores (mazaedium) borne on delicate stipes or adnate on the thallus (see fig. 66b).
    10. Stipes present. . . . . *Coniocybe*
    10. Mazaedia adnate on the thallus. . . . *Cyphelium*
  9. Hymenium with vertically arranged paraphyses (deliquescing only in *Verrucaria*) and asci containing spores; fruiting bodies not stiped.
    11. Fruiting bodies appearing as straight or wavy black lines. *Graphis*
    11. Fruiting bodies round and disk-shaped or immersed in the thallus, whitish to pale brown or black.
      12. Fruiting bodies perithecia, immersed in the medulla and appearing as black dots on the surface (fig. 13b). . . . . *Anisomeridium*
      12. Fruiting bodies apothecia, adnate on the thallus or immersed, but with an open disk (pored only in *Pertusaria*).
        13. Spores brown.
          14. Thalline rim surrounding the disk; rim of the same color as the thallus. . . . . *Rinodina*
          14. Thalline rim lacking; disk with or without a raised black rim.
            15. Thallus consisting of convex areoles separated by black lines (see fig. 71d, pl. 9c).
            . . . . *Rhizocarpon*

15. Thallus continuous and smooth, without areoles.
    16. Apothecia flush with the thallus surface, appearing like craters. . . . . *Diploschistes*
    16. Apothecia adnate, raised with an open disk. . . . . *Buellia*
13. Spores colorless.
    17. Spores muriform.
        18. Apothecia becoming buried under sorediate masses. . . *Phlyctis*
        18. Apothecia not sorediate. *Thelotrema*
    17. Spores simple to transversely septate.
        19. Spores simple, without cross walls.
            20. Spores more than 30 μm long (often reaching 80–90 μm).
                21. Apothecia immersed in warts, opening through small pores. . . . . *Pertusaria*
                21. Apothecia flush with the thallus surface or adnate to emergent; disk open.
                    22. Apothecia flush to somewhat sunken; disk concave, about 1 mm in diameter. . . . . *Pachyospora*
                    22. Apothecia adnate to emergent, large, 2 mm or more in diameter.
                        23. Thalline rim well devel-

oped; disk pinkish tan. . .
Ochrolechia

23. Thalline rim lacking; disk black. . . . . . .
Mycoblastus

20. Spores less than 30 μm long (often only 10–20 μm long).

24. Thallus yellow to lemon yellow. . . . .
. . . . *Candelariella*

24. Thallus greenish or whitish gray to brown.

25. Apothecia with a distinct thalline rim of the same color as the thallus. . . . .
. . . . *Lecanora*

25. Apothecia lacking a thalline rim (a raised black rim may or may not be present).

26. Apothecia adnate or raised with a flat to convex disk.
. . . . *Lecidea*

26. Apothecia flush to sunken with a concave disk. . . . .
. . . . *Aspicilia*

19. Spores transversely septate with one or more cross walls.

27. Thalline rim lacking; disk

with or without a raised
proper rim. . . . . . . *Bacidia*
27. Thalline rim well developed.
28. Disk deep red, K+ purple. . . . . *Haematomma*
28. Disk whitish, K−. . . .
. . . . *Roccellina*

Acarospora Mass.

Acarospora is a widespread crustose genus occurring mostly
on exposed acidic rocks in sunny locations from the deserts to
high mountains. It is easily recognized in the field by the
chinky, areolate thallus with immersed apothecia opening by a
small or expanded pore. The asci contain numerous tiny spores.
Taxonomically, however, the genus is poorly understood and
many of the 34 species reported from California are doubtfully
identified. The two species included here are unquestionably
the most common representatives in the state.

1. Thallus greenish to lemon yellow. . . . . . . *A. chlorophana*
1. Thallus dark brown. . . . . *A. fuscata*

Acarospora chlorophana (Wahlb.) Ach.   (pl. 1a)

Thallus bright lemon to greenish yellow, very closely adnate,
2–10 cm broad but often fusing into larger colonies, the margins more or less lobate and the surface chinky-areolate. Apothecia common, to 1 mm wide, disk pale brownish green, partially enclosed by thalline tissue; spores colorless, simple, 32
to 64 per ascus, 1–2 μm. Thallus K+ yellowish (rhizocarpic
acid; medullary components not determined).

Habitats: On exposed acidic rocks (usually on vertical faces)
from deserts into Montane and Subalpine Forest from near sea
level to 11,000 ft elevation.

Range: See fig. 72a.
This species is typical of a poorly understood group of

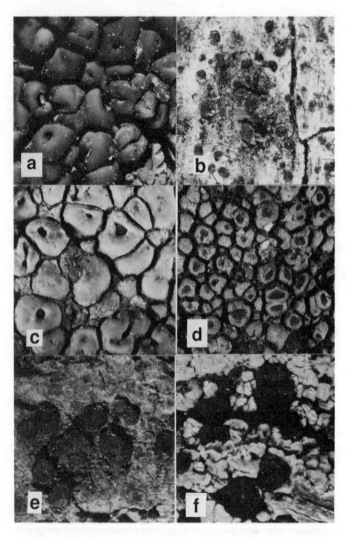

FIG. 65. Species of crustose lichens: (a) *Acarospora fuscata;*
(b) *Anisomeridium biforme;* (c) *Aspicilia cinerea;* (d) *A. caesiocinerea;* (e) *Bacidia laurocerasi;* (f) *Buellia disciformis* (all ×10).

yellow Acarosporas (see Weber, 1968). A second common species, *A. schleicheri,* consists of scattered areoles without marginal lobation; it is the commoner form on flat rocks. Both species can be told from yellow *Rhizocarpon geographicum* by the lack of a black hypothallus.

Acarospora fuscata (Nyl.) Arn.   (fig. 65a)

Thallus dark to pale brown, weakly lobed at the margins with a chinky-areolate center, 2–4 cm broad. Apothecia immersed in the areoles, expanding at maturity with an open, brown disk to 0.5 mm in diameter; spores simple, colorless, 32 to 64 per ascus, 2×3 μm. Thallus K−; medulla K− and C+ rose (gyrophoric acid).

Habitats: On more or less exposed acidic rocks (gneiss, granite, slate) in Valley and Foothill Woodland into Subalpine Fell-Fields at 300 to 10,000 ft elevation.

Range: Rather common from San Diego County northward to Glenn County in the North and South Coast Ranges and from Kern County to Mono County in the Sierra Nevada.

The C+ color test is usually sufficient for identifying this common species. Brown specimens with a C− reaction are also found, but our knowledge of their taxonomy is too poor to give names to the species.

Anisomeridium (Müll. Arg.) Choisy

Anisomeridium is typical of pyrenocarpous lichens ("pyrenocarps"), a very large group of poorly known crustose species belonging to several different families and intergrading in many cases with the true fungi. The fruiting bodies, called perithecia, are flask-shaped, more or less immersed in the thallus, and open by a tiny pore. They are usually small, black, and brittle and very difficult to section and study. The genera are recognized by various combinations of paraphysis structure, nature of the perithecial wall, and spore septation and color. The following keys include genera reported from California, but it is impossible to go into more detail until taxonomic revisions are available.

[1. Paraphyses gelatinized and disappearing, only asci remaining distinct.................... Verrucariaceae
(*Polyblastia, Trimmatothele, Verrucaria*)]
1. Paraphyses persistent.
  [2. Ascocarps with more than one chamber...........
  .... *Tomasellia*]
  2. Ascocarps with one chamber.
    [3. Ascocarp wall blue-green............. *Porina*]
    3. Ascocarp wall black or pale.
      [4. More than eight spores per ascus.........
      .... *Thelopsis*]
      4. Eight or fewer spores per ascus.
        [5. Phycobiont blue-green; collected on marine rocks........... *Pyrenocollema*]
        5. Phycobiont green; collected on trees or rocks.
          [6. Paraphyses unbranched. ... *Porina*]
          6. Paraphyses branched and netlike.
            7. Spores two-celled; septum eccentric........... *Anisomeridium*
            [7. Spores two or more celled; septum transverse...............
            .... *Arthopyrenia*]

Anisomeridium biforme (**Borr.**) **Harris**   (fig. 65b)

Thallus whitish, smooth, thin, 2–6 cm broad. Perithecia numerous, partially immersed, very tiny, to 0.2 mm in diameter, black and carbonaceous; paraphyses branched; spores colorless, two-celled, 4–6 × 8–12 μm. Thallus K–, C–, P– (no substances present).

Habitats: On oak trees in Valley and Foothill Woodland from near sea level to 1000 ft elevation.

Range: San Diego County north to the San Francisco Bay area in the North and South Coast Ranges.

Aspicilia Mass.

This genus is probably the commonest saxicolous crust in California, occurring over a wide range of habitats. It has a rather thick, chinky but continuous crustose thallus with flush to sunken black apothecia, which lack a rim. The spores are colorless and simple. This genus was formerly placed in *Lecanora,* which has raised apothecia with a distinct thalline rim. About 10 species are known in California, but none of these has been revised and in fact little is known of the genus in North America. *Aspicilia cinerea,* reacting K+ red, is often accompanied on the same rocks with *A. caesiocinerea* (Nyl.) Arn. (fig. 65d), which reacts K− (no substances present) and has somewhat smaller apothecia. *Aspicilia calcarea* (L.) Mudd, also very similar externally, has two to four spores per ascus.

Aspicilia cinerea (L.) Krb. (fig. 65c)

Thallus light mineral to pale brownish gray, chinky-areolate without marginal lobes, 4–10 cm broad. Apothecia numerous, at first black dots but expanding to form a black angular disk with a slightly raised, often pruinose rim; spores simple, colorless, eight per ascus, 4–5 × 8–12 μm. Thallus K+ yellow turning red, C−, P+ orange (norstictic acid).

Habitats: On more or less sheltered outcrops of gneiss, granite, and other acidic rocks in Valley and Foothill Woodland and Montane Forest at 1000 to 6000 ft elevation.

Range: Very common from San Diego County northward to Tehama County in the North and South Coast Ranges and from Kern County to Butte County in the Sierra Nevada.

Bacidia De Not.

This crustose genus is a member of the family Lecideaceae. It is representative of those genera with colorless, transversely septate spores. The species are widespread, with 23 reported from California, occurring over mosses and on bark and rocks, but without modern revisions it is well-nigh impossible to identify the specimens accurately. As with other genera in this large family, the apothecia must be carefully sectioned and examined under a microscope for spore characters. Other confusable genera with septate spores include *Catillaria, Toninia,* and *Lecanactis. Catillaria* has two-celled spores but is otherwise very close to *Bacidia. Toninia,* a widespread lichen on soil, rocks, and the base of trees, has a well-developed, thick, and bullate thallus. *Lecanactis* is actually classified in the family Lecanactidaceae with *Schismatomma,* but because of the *Bacidia*-like apothecia it matches the key for *Bacidia* here.

[1. Thallus well developed, bullate to subsquamulose. . . . . . .
. . . . *Toninia*]
 1. Thallus thin, continuous, and plane.
   [2. Spores ovoid, two-celled. . . . . *Catillaria*]
   2. Spores long ellipsoidal to acicular, three–seven-
     septate.
     [3. Spores 6 × 20 μm, three-septate; paraphyses
       branched. . . . . *Lecanactis*]
     3. Spores 2 × 30–40 μm, three–seven-septate,
      acicular; paraphyses unbranched. . . . . . . *Bacidia*

Bacidia laurocerasi (Duby) Ozenda & Clauz.   (fig. 65e)

Thallus crustose, thin, greenish to brownish gray, 2–5 cm broad. Apothecia numerous, sessile, 0.5–1.2 mm in diameter, disk dark, flat or becoming convex at maturity, the proper rim dark; spores colorless, transversely septate with seven to nine locules, 3–6 × 20–24 μm, eight per ascus. Thallus K– (chemistry not determined).

Habitats: On trunk and branches of broadleaf trees and conifers in Valley and Foothill Woodland from sea level to 2000 ft elevation.

Range: Widespread from Los Angeles County northward to the Bay Area in the Coast Ranges.

Buellia De Not.

Buellia is a large genus of rock and bark-inhabiting lichens. The spores are brown and two-celled to rarely muriform, and the apothecia lack a thalline rim. Externally *Lecidea, Lecidella, Rhizocarpon,* and other genera in the Lecideaceae are very similar but have different spores. A monograph by Imshaug (1951) lists 22 species for California, most in the drier, low-elevation forests from the Bay Area south to San Diego County. About eight of them grow only on rocks while the remainder are found on trees, dead branches, and fenceposts. The species are told apart by various microscopic characters, color of the hypothecium and epithecium, spore septation and size, and thickness of spore walls. Imshaug's monograph should be consulted by anyone attempting to identify material to the species level. The commonest group is centered around *B. parasema–B. oidalea–B. penichra,* all common in oak forests, growing intermixed with *Lecidella, Rinodina,* and other crusts.

Buellia disciformis (Fr.) Mudd (fig. 65f)

Thallus whitish to tannish gray, 2–6 cm broad; surface smooth to becoming bullate or verruculose. Apothecia common, to 1 mm in diameter, proper rim more or less visible, disk black, becoming convex; spores brown, two-celled, 7–15 × 20–32 μm. Thallus surface K+ yellow turning red, P+ orange (norstictic acid).

Habitats: Oaks, pines, *Artemisia,* and other trees in Valley and Foothill Woodland and Montane Forest from near sea level to 3000 ft elevation.

Range: San Diego County north to Shasta County in the North and South Coast Ranges, Cascades, and Sierra Nevada.

Caloplaca Th. Fr.

As presently conceived, *Caloplaca* is a large, heterogeneous group of primarily crustose species. Also included, however, are a few marginally lobate species such as *C. saxicola.* There is even a small fruticose species, *C. coralloides* (Tuck.) Hult., which occurs on rocks above the spray zone from the Channel Islands to Sonoma County. It is very similar to *Edrudia constipans* (Nyl.) Jordan, which differs in having simple spores; it is known only from the Farallon Islands.

All species of *Caloplaca* have an orange apothecial disk which reacts K+ purple. Thalloid or vegetative structures may also be orange and react with K. The spores are colorless, two-celled or polarilocular. The taxonomy of the genus is poorly known at this time. The 25 to 30 crustose species reported from California are separated by difficult and intergrading hymenial and spore characters. They occur very widely on trees and on mostly calcareous rocks. *Caloplaca saxicola* is common and one of the first crustose lichens collected by beginners.

1. Thallus entirely crustose with no marginal lobation. . . . . .
. . . . *C. cerina*
1. Thallus crustose with marginal lobation or minutely fruti-cose.
    [2. Thallus minutely fruticose. . . . . *C. coralloides*]
   2. Thallus crustose with marginal lobation.
       [3. Lower surface easily freed from rock; cortex and sparse rhizines present. . . . . . *Xanthoria elegans*]
      3. Lower surface appressed to rock; cortex lacking.
         [4. Lobes with marginal soredia. . . . .
. . . . *C. cirrochroa*]
        4. Lobes lacking soredia.
           [5. Thallus mostly areolate with weakly devel-oped marginal lobation. . . . . *C. lobulata*]
          5. Thallus thick and chinky with well-devel-oped, crowded marginal lobes.
             6. Lobes flattened to convex, smooth; found throughout the state. . . . .
. . . . *C. saxicola*
             6. Lobes strongly convex with a warty surface; found mostly in desert areas.
. . . . *C. trachyphylla*

Caloplaca cerina (Ehrh.) Th. Fr.   (fig. 8)

Thallus crustose, minutely warty and sometimes lacking in part, whitish gray, 1–4 cm broad. Apothecia abundant, be-coming crowded, 0.4–0.8 mm in diameter, disk deep rusty orange with a distinct light orange rim; spores colorless, polarilocular, $5-8 \times 12-18$ μm. Thallus and apothecial disk K+ purple (parietin).

Habitats: On trunk and branches of broadleaf trees (oaks, *Aesculus, Heteromeles,* and others) and conifers in Valley and Foothill Woodland into Montane Forest from sea level to 7000 ft elevation.

Range: Common from San Diego County northward to Glenn County in the North and South Coast Ranges and from Kern County to Plumas County on the western slopes of the Sierra Nevada.

This species is representative of the many purely crustose species of *Caloplaca.*

Caloplaca saxicola (Hoffm.) Nordin (pl. 1a)

Thallus orange to deep orange red, closely appressed, up to 2 cm broad, fusing into larger colonies; marginal lobes distinct, crowded, 0.5–1.5 mm wide, becoming somewhat convex, central part of thallus usually chinky-areolate; lower surface where exposed white, cortex lacking, rhizines lacking. Apothecia very common, to 1.5 mm wide, disk orange with a somewhat darker rim; spores colorless, two-celled, $5-6 \times 14-15$ μm. Cortex K+ purple; medulla K−, C−, P− (parietin in cortex).

Habitats: On exposed rock outcrops in Valley and Foothill Woodland, Montane Forest, and Subalpine Forest from near sea level to 7500 ft elevation.

Range: See fig. 72b.

This is one of the most easily recognized lichens in California because of the deep orange-red color. Colonies may fuse together to cover large areas of rock. *Xanthoria elegans,* a similar but rarer orange lichen, has a lower cortex and can be removed from rocks without as much damage.

There are two other lobate Caloplacas in California. One has tiny sorediate lobes: *C. cirrochroa* (Ach.) Th. Fr., which is rather rare but widespread on lava outcrops. The other species, *C. lobulata* (Flk.) Hellb., lacks soredia and could be confused with *C. saxicola*. It has smaller apothecia and dispersed areoles without distinct marginal lobes. It is very common on rocks in Valley and Foothill Woodland.

Caloplaca trachyphylla (Tuck.) Zahlbr. (pl. 1d)

Thallus orange to salmon-orange, closely adnate, 1–3 cm broad, fusing into larger colonies; lobes 0.5–1 mm wide, rather long, convex, and appearing warty with age; lower surface where exposed light orange, cortex and rhizines lacking. Apothecia common, 0.5–1 mm in diameter, disk orange with a salmon-orange rim; spores 6 × 12 μm. Cortex K+ purple; medulla K−, C−, P− (parietin).

Habitats: On rocks in open, semiarid areas and in Montane and Subalpine Forest from near sea level to 9000 ft elevation.

Range: Widespread in Imperial County and from Kern County to Modoc County in the Sierra Nevada, Cascades, and Modoc Plateau.

Candelariella Müll. Arg.

This minutely squamulose crustose genus is very widespread on rocks in California and easily recognized in the field by the bright yellow thallus even when scraps are collected. The genus is told from orange *Caloplaca* by the K− thallus reaction. Six species are known in California, but only *C. rosulans* and *C. vitellina* (Hoffm.) Müll. Arg. are commonly collected. Number of spores per ascus is important but may be rather difficult to determine since the apothecia are tiny and spores often poorly developed.

1. Thallus consisting of large rosette-forming squamules. . . .
   . . . . *C. rosulans*
1. Thallus crustose, consisting of small continuous or dispersed squamules.
   [2. Eight spores per ascus; squamules dispersed. . . . .
   . . . . *C. aurella*]

[2. Twelve or more spores per ascus; squamules crowded.
.... *C. vitellina*]

Candelariella rosulans (Müll. Arg.) Zahlbr.   (pl. 2b)

Thallus light cadmium yellow to lemon yellow, very closely adnate, consisting of small lobulate squamules fused into rosettes, 1–5 cm broad. Apothecia about 0.5 mm wide, disk yellow with an entire to minutely crenate rim; spores colorless, simple, 5 × 15–18 µm, eight per ascus. Thallus K− (calycin).

Habitats: On granite, gneiss, and sandstone in the Valley and Foothill Woodland into Montane and Subalpine Forest from 1000 to 9000 ft elevation.

Range: Imperial and San Diego counties northward to the San Francisco Bay area in the South Coast Ranges and from Kern County to Modoc County in the Sierra Nevada and Modoc Plateau.

Catapyrenium Flot.

This genus of dark brown, squamulose lichens was formerly classified in *Dermatocarpon*, which now includes only umbilicate species. Both genera have immersed perithecia. *Catapyrenium* typically occurs on soil in semiarid or desert regions and acts as a soil consolidator. It can, however, be confused with *Psora*, which has disk-shaped apothecia, or even with some Peltulas, which have flush apothecia; both of these genera also grow on soil. A monograph by Thomson (1987) discusses the six species known in California. *Catapyrenium lachneum* is the commonest of these.

Catapyrenium lachneum (Ach.) Sant.   (fig. 66a)

Thallus dark brown, consisting of flattened, appressed squamules 2–5 mm long, the edges barely ascending, forming colonies 4–10 cm broad; lower surface dark, bare. Perithecia immersed in the thallus and opening by a tiny black pore; spores simple, colorless, 6–9 × 10–13 µm; pycnidia abundant, resembling the perithecia but containing a mass of tiny conidia. Medulla K−, C− (no substances present).

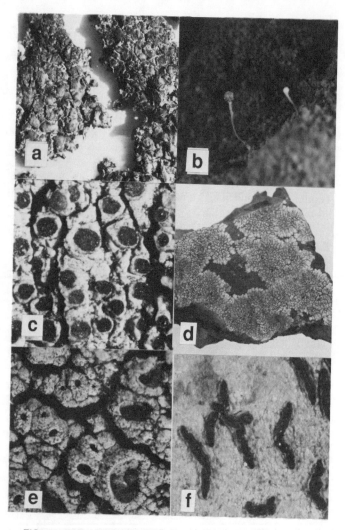

FIG. 66. Species of crustose lichens: (a) *Catapyrenium lach-neum;* (b) *Coniocybe furfuracea;* (c) *Cyphelium tigillare;* (d) *Di-melaena radiata;* (e) *Diploschistes scruposus;* (f) *Graphis scripta* (d: ×1; a–c, e, f: ×10).

Habitats: On soil, soil in rock crevices, and over rocks in open areas in Valley and Foothill Woodland and deserts from sea level to 3000 ft elevation.

Range: Widespread but often overlooked from San Diego County northward into central California in the South Coast Ranges, western foothills of the Sierra Nevada, and the Mojave Desert.

Coniocybe Ach.

Coniocybe is a typical member of the order Caliciales. This order (excluding *Sphaerophorus*) consists of crustose lichens with unique fruiting bodies. They resemble black apothecia, but the hymenium disintegrates into a mass of loose spores called a mazaedium. There are two broad groups, one like *Coniocybe* with delicate, funguslike stalked apothecia only 1–2 mm high and one with adnate apothecia, as in *Cyphelium* (see the following description). The species often grow on fenceposts and at the base of conifers (especially cypress). They are easily overlooked and require special care in curation to prevent breaking and loss of the stalks. See Tibell (1975) for more details.

[1. Spores one-septate. . . . . *Calicium*]
  1. Spores simple, one-celled.
      2. Spores pale brown to colorless. . . . . *Coniocybe*
      2. Spores dark brown.
          [3. Spores spherical. . . . . *Chaenotheca*]
          [3. Spores ellipsoidal. . . . . *Mycocalicium*]

Coniocybe furfuracea (L.) Ach.   (fig. 66b)

Thallus crustose, powdery, yellow-green, 3–8 cm broad. Apothecia black, borne on delicate stalks 1–2 mm high; spores massed in the apothecia, brown, simple, $2 \times 3 \ \mu m$. Thallus K− (rhizocarpic and vulpinic acids).

Habitats: Base of conifers and on stumps and fences in mature forests in North Coastal Forest from near sea level to 1000 ft elevation.

Range: Widespread from Monterey County northward to Del Norte County in the North and South Coast Ranges and in Nevada County on the western slopes of the Sierra Nevada.

Cyphelium Ach.

This member of the order Caliciales (see *Coniocybe* above) has a well-developed crustose thallus and large, sessile, black apothecia. As in *Coniocybe,* the disk consists of a mass of dark, loose spores called a mazaedium. The most widespread species, *C. tigillare,* has a deep yellow thallus and occurs on fenceposts. A related genus, *Thelomma,* is represented in California by *T. mammosum* (Hepp) Mass. (*Cypheliopsis bolanderi*), a conspicuous gray crust with single-celled brown spores. *Texosporium* has an orange-yellow layer below the hymenium and unusual roughened, two-celled spores. Species recently segregated in the genus *Thelomma* are discussed in detail by Tibell (1976), and additional keys can be found in Weber (1967).

[1. Spores large, 30–35 μm long, the wall strongly roughened; base of mazaedium orange-yellow. . . . .
. . . . *Texosporium sancti-jacobi*]
1. Spores smaller, less than 20 μm long; mazaedia lacking pigments.
    [2. Spores simple. . . . . *Thelomma mammosum*]
    2. Spores one-septate.
        3. Thallus deep golden yellow. . . . .
        . . . . *Cyphelium tigillare*
        3. Thallus gray to pale yellowish brown.
            4. Spores at maturity with striations on walls; exciple black. . . . . *Cyphelium*
            [4. Spores with a network of minute ridges on walls; exciple colorless. . . . . *Thelomma*]

Cyphelium tigillare (Ach.) Ach.   (fig. 66c)

Thallus crustose, warty, and verrucose, yellow-green, 3–10 cm broad. Apothecia common, immersed to raised on the warts, to 0.8 mm in diameter, disk jet black, dull; spores massed in the mazaedium, brown, septate, 10 × 15 μm. Thallus K−, C−, P− (rhizocarpic acid).

Habitats: On fenceposts, dead wood, and base of conifers in Valley and Foothill Woodland and Montane Forest from sea level to 6000 ft elevation.

Range: Rather common from Los Angeles County northward to the San Francisco Bay area in the South Coast Ranges.

Dimelaena Norm.

This subcrustose, marginally lobate genus might be mistaken at first for a small foliose lichen, but the lower surface lacks a cortex and is fused to the rock substratum. The apothecia have a thalline rim; the brown, two-celled spores are easily observed under the microscope. The only yellow-green lichen that could be confused with *Dimelaena* is *Lecanora muralis,* which has colorless, simple spores. *Dimelaena* species prefer large smooth granitic boulders and are often difficult to collect.

Of the four species of *Dimelaena* reported in California, two (*D. californica* (Magn.) Sheard and *D. thysanota* (Tuck.) Hale & Culb.) have only been rarely collected in the South Coast Ranges. They are dark brown with only weak development of marginal lobules in comparison with *D. oreina* and *D. radiata* and might well be confused with *Buellia* or *Rinodina* species. See Sheard (1974) for the most recent monographic study.

1. Thallus yellow-green. . . . . *D. oreina*
1. Thallus whitish gray to dark brown.
   2. Thallus whitish gray. . . . . *D. radiata*
   2. Thallus brown.
      [3. Medulla K+ red (norstictic acid). . . . .
      . . . . *D. californica*]
      [3. Medulla K−. . . . . *D. thysanota*]

Dimelaena oreina (Ach.) Norm.   (pl. 3c)

Thallus yellow-green, closely appressed, 2–6 cm broad; marginal lobes distinct, 1 mm wide, the central part becoming chinky and brownish; lower surface fused with rock, lacking rhizines. Apothecia very common, to 1 mm in diameter, disk black with a distinct yellow rim; spores brown, two-celled, 5 × 10 μm. Cortex K−; medulla K−, C+ red, P− or K+, P+

yellow, C− (usnic acid and either gyrophoric acid and sphae-rophorin or stictic acid).

Habitats: On open rock outcrops (especially granite) in Valley and Foothill Woodland and Montane Forest from 500 to 8000 ft elevation.

Range: See fig. 73b.

The chemical population with sphaerophorin occurs in the Coast Ranges and Sierra Nevada foothills below 4000 ft elevation. The stictic acid population is collected only above 5000 ft in the San Bernardino Mountains and the Sierra Nevada.

Dimelaena radiata (Tuck.) Hale & Culb. (fig. 66d)

Thallus whitish to brownish or pale yellowish gray, closely appressed, 2−4 cm broad but fusing into larger colonies; marginal lobes distinct, about 0.5 mm wide, the central part of the thallus chinky; lower surface fused to the rock, rhizines lacking. Apothecia very common, to 0.5 mm in diameter, disk black or barely white pruinose, rim whitish gray; spores brown, two-celled, 5 × 9 μm. Cortex and medulla K−, C−, P− (unknown substances).

Habitats: Open rock outcrops in Valley and Foothill Woodland and Coastal Scrub near sea level to 2000 ft elevation.

Range: Widespread from San Diego County northward to the San Francisco Bay area in the South Coast Ranges and Tulare County in the Sierra Nevada.

Diploschistes Norm.

This crustose genus is very common on rocks in California. It is recognized by the thick whitish thallus and the large, more or less sunken apothecia with brown, muriform spores. It is actually related to *Thelotrema lepadinum,* which grows on trees, and they are both classified in the family Thelotremataceae.

1. Thallus thick with apothecia to 1 mm in diameter. . . . . . .
. . . . *D. scruposus*

[1. Thallus thin with numerous small apothecia less than 0.5 mm in diameter. . . . . *D. actinostomus*]

Diploschistes scruposus (Schreb.) Norm.   (fig. 66e)

Thallus forming a thick whitish gray crust, closely adnate, chinky toward the center, up to 10 cm broad. Apothecia common, sunken to adnate on the thallus, 0.5–1 mm in diameter, disk black with a raised rim partially free of the disk; spores brown, muriform, $15 \times 30 \ \mu$m, four per ascus. Thallus K−, C+ red, P− (lecanoric acid).

Habitats: On open rocks or in part on soil in North Coastal and Montane Forest and Valley and Foothill Woodland from 1000 to 4000 ft elevation.

Range: See fig. 73c.
 A related species, *D. actinostomus* (Pers.) Zahlbr., is much smaller. It occurs more rarely from San Diego to Tehama County in the North and South Coast Ranges and in Madera County on the western slopes of the Sierra Nevada.

Graphis Adans.

No other group of lichens is more easily recognized in the field than the family Graphidaceae. The black fruiting bodies resemble scratches or writing on tree bark. These linear apothecia are called lirellae. Three closely related genera in California are separated by spore color and septation: *Graphina*, *Graphis*, and *Phaeographis*. Some other lirelliform genera are externally similar but are classified in other families: *Opegrapha* (Opegraphaceae) has branched paraphyses; *Xylographa* (Agyriaceae) has simple spores and lacks a distinct thallus; and *Arthonia* (Arthoniaceae) has rather indistinct, irregularly shaped, flush lirellae with an ascolocular hymenium and one–three-septate spores with unequal cells. All lirelliform specimens must be sectioned and examined under a microscope for positive identification.

[1. Spores brown. . . . . *Phaeographis*]
 1. Spores colorless.
     [2. Spores muriform. . . . . *Graphina*]

2. Spores simple or transversely septate.
   [3. Spores simple; thallus not visible on substrate. . . .
   . . . . *Xylographa*]
  3. Spores transversely septate; thallus usually distinctly developed.
   [4. Apothecia irregularly shaped and flush with the thallus. . . . . *Arthonia*]
   4. Apothecia sharply defined and linear, raised.
     5. Paraphyses unbranched. . . . . *Graphis*
     [5. Paraphyses branched. . . . . . *Opegrapha*]

Graphis scripta (L.) Ach.   (fig. 66f)

Thallus crustose, thin and smooth, whitish gray, 2–5 cm broad. Apothecia black, linear, about 0.2 mm wide and 2–3 mm long, straight to curved and branched; paraphyses simple; spores colorless, six–nine-septate, $8–12 \times 30–50$ μm. Thallus K− (no substances present).

Habitats: On twigs and trunks of oaks and other broadleaf trees (alder, Tan Bark Oak, and others) in mature forests in the North Coastal Forest from near sea level to 2000 ft elevation.

Range: Widespread from the San Francisco Bay area northward to Humboldt County in the North Coast Ranges.

Haematomma Mass.

This crustose genus, a member of the family Lecanoraceae, is unique in having a red disk which reacts with KOH. It is representative of several other rare or overlooked genera in the family in California, which have transversely septate spores. For example, *Icmadophila ericetorum* (L.) Zahlbr., also classified in the Lecanoraceae, has conspicuous pinkish, flesh-colored apothecia and grows on rotted logs in Humboldt County. *Dimerella lutea* (Dicks.) Trev. is an inconspicuous lichen with a pale orange-brown disk, growing over mosses on tree trunks. Both *Dirina* Fr. and *Roccellina* (recently revised by Tehler, 1983) are *Lecanora*-like but have ascolocular hymenia; they occur in southern coastal areas.

1. Disk deep rusty red. . . . . *Haematomma pacificum*
1. Disk whitish to orangy or flesh-colored.
    [2. Spores one-septate; apothecia delicate, less than 1 mm wide. . . . . *Dimerella lutea*]
    2. Spores one to many septate; apothecia large, 1–2 mm wide.
        [3. Collected on rotted logs in North Coastal Forest; disk pink. . . . . *Icmadophila ericetorum*]
        [3. Collected on trees and rocks in southern coastal California; disk white. . . . . *Roccellina*]

Haematomma pacificum **Hasse** (fig. 67a)

Thallus continuous to verrucose and subareolate, whitish gray, 4–7 cm broad. Apothecia conspicuous, 1–2 mm in diameter, disk deep rusty red, surrounded by a thin whitish rim; spores colorless, four–seven-septate, 2–3 × 30–40 μm. Thallus K–, C–, P– (substances not determined).

Habitats: On broadleaf trees, conifers, and dead wood in open forests in North Coastal Forest and Valley and Foothill Woodland from 2000 to 4000 ft elevation.

Range: Widespread from Los Angeles County northward to Mendocino County in the North and South Coast Ranges and in Yuba County in the Sierra Nevada foothills.

Hypocenomyce Choisy

This is a typical squamulose-crustose genus, the thallus consisting of crowded, suberect squamules. It has usually been classified in *Lecidea* or *Psora*. The most frequent habitat is charred stumps where it forms an extensive greenish crust.

Hypocenomyce scalaris **(Ach.) Choisy** (fig. 67b)

Squamules light green to brownish, closely adnate, up to 1 mm wide but forming colonies 5–20 cm broad, the margins suberect, smooth to dissected and sorediate; lower surface white, lacking a cortex. Apothecia absent. Thallus K–, C+ red, P– (lecanoric acid).

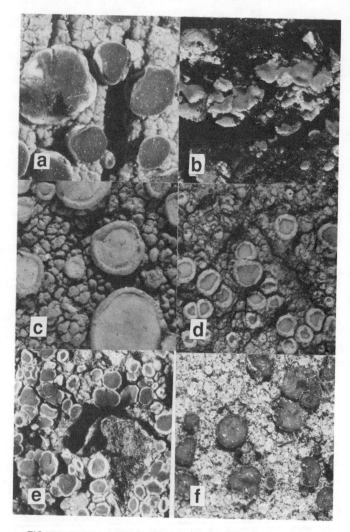

FIG. 67. Species of crustose lichens: (a) *Haematomma pacificum*; (b) *Hypocenomyce scalaris*; (c) *Lecanora caesiorubella*; (d) *L. pacifica*; (e) *L. varia*; (f) *Lecidea auriculata* (all ×10).

Habitats: On charred stumps in North Coastal and Montane Forest from 2000 to 6000 ft elevation.

Range: Widespread (but often overlooked) in Santa Barbara County and the San Francisco Bay area and northward in the North Coast Ranges and from Fresno County northward on the western slopes of the Sierra Nevada.

The crusty squamules are much smaller than those of most Cladonias and can be easily identified by the C+ red reaction.

Lecanora Ach.

Lecanora, with nearly 1000 species, is one of the largest lichen genera. About 60 species have been reported from California, although some of these are now classified in segregate genera (see *Aspicilia* and *Rhizoplaca*). All species have a thalline rim around the apothecial disk and simple, colorless spores. Identification of the species is based mostly on spore and hymenial characters which require detailed microscopic study. Since there are no recent revisions of the taxonomy of the western American species, neither the professional nor the amateur lichenologist can identify all of them with confidence. The five species presented here are probably the most commonly collected ones and representative of the genus. Keys published by Weber (1963) for Arizona and by Wetmore (1967) for South Dakota can also be used to identify California species, along with Brodo's (1984) recent revision of the common *Lecanora subfusca* group.

1. Thallus crustose but chinky at the center with distinct marginal lobation.
   2. Thallus greenish yellow. . . . . *L. muralis*
   2. Thallus cinnamon brown. . . . . *L. mellea*
1. Thallus uniformly crustose without marginal lobes.
   3. Apothecia large and conspicuous, 1–3 mm in diameter.
      4. Thallus K+ red, C−; spores 14–16 μm long. . . .
      . . . . *L. caesiorubella*
     [4. Thallus K−, C+ pink; spores 50–80 μm long. . . .
     . . . . *Ochrolechia pallescens*]
   3. Apothecia small, not exceeding 1 mm in diameter.

5. Apothecial disk brown with a thick whitish thalline rim; thallus well developed. . . . . *L. pacifica*

5. Apothecial disk pale greenish yellow with a thin rim; thallus thin and granular. . . . . *L. varia*

Lecanora caesiorubella Ach. (fig. 67c)

Thallus crustose, continuous, minutely warty, pale greenish gray, 2–8 cm broad. Apothecia large and conspicuous, 1–2 mm in diameter, disk pale tannish yellow or in part white pruinose, the greenish gray rim usually thick and well developed; spores simple, colorless, 6–10 × 14–16 μm, eight per ascus. Thallus K+ yellow, C–, P+ orange (atranorin, norstictic, and protocetraric acids).

Habitats: On trunks and branches of oak and other broadleaf trees (alder, maple, Tan Oak) in mature forests in North Coastal Forest and Valley and Foothill Woodland from near sea level to 1500 ft elevation.

Range: Fairly common from San Diego County northward to the San Francisco Bay area and Humboldt County in the North and South Coast Ranges.

This species could be confused with *Ochrolechia pallescens,* which has much larger spores and reacts K–, C+ red.

Lecanora mellea W. Web. (pl. 4d)

Thallus crustose, shiny, orange to cinnamon brown, with crenate margins and lobes to 1 mm wide, the colonies 2–5 cm broad. Apothecia common, 1–2 mm in diameter, disk brownish orange with a broad rim of nearly the same color; spores simple, colorless, 6–8 × 9–11 μm, eight per ascus.

Habitats: On sheltered rocks (granite, gneiss, and others) in North Coastal and Montane Forest and Valley and Foothill Woodland from 500 to 7000 ft elevation.

Range: See fig. 74d.

Lecanora muralis (Schreb.) **Rabenh.**   (pl. 5a)

Thallus greenish to pale yellow, 3–5 cm wide and often fusing
into larger colonies; lobes linear, separate to crowded, 0.5–1
mm wide, chinky with age, the margins black-rimmed or white
pruinose, the tips blackening and sometimes free; lower surface in close contact with rock, cortex lacking. Apothecia numerous, crowded, 0.5–1.3 mm wide, disk tan to dark brown,
the rim becoming white pruinose; spores 6–8 × 11–15 μm,
eight per ascus. Cortex and medulla K–, C–, P– (usnic acid
and zeorin).

Habitats: On sheltered rocks in North Coastal and Montane
Forest, Valley and Foothill Woodland, and Subalpine Forest
from near sea level to 9000 ft elevation.

Range: Very common throughout the state.

This is one of the commonest rock lichens in California.
Because of the yellow-green thallus it could be mistaken for
Xanthoparmelia species, which differ in having a distinct
lower cortex and rhizines. Also, superficially similar *Dimelaena oreina* has brown, two-celled spores.

Lecanora pacifica **Tuck.**   (fig. 67d)

Thallus crustose, smooth to minutely warty, whitish gray, 1–2
cm broad. Apothecia numerous, 0.5–1.5 mm in diameter, disk
dark to pale brown with a distinct whitish gray rim; spores
simple, colorless, 8 × 15 μm, eight per ascus. Thallus K+,
P+ yellow, C– (atranorin).

Habitats: On trunks and branches of oaks, *Myrica,* maple,
and other broadleaf trees in North Coastal and Montane Forest
and Valley and Foothill Woodland from near sea level to 6000
ft elevation.

Range: Common from Riverside County northward to Mendocino County in the North and South Coast Ranges and
Tulare County to Shasta County in the Sierra Nevada.

In the older literature there are many references to "*Lecanora subfusca.*" This name stands for a widespread crustose species complex which has a distinct rim around the pale brown disk and simple colorless spores. Brodo (1984) recently revised this difficult group in North America and discovered at least 14 different species in California, separated by various apothecial characters requiring careful microscopic examination. The commonest species seems to be *L. pacifica.* Anyone interested in more details should consult Brodo's article.

Lecanora varia (Hoffm.) Ach.

(fig. 67e)

Thallus crustose, poorly developed, granular, sometimes lacking, yellowish green, 1–5 cm broad. Apothecia numerous and crowded at maturity, 0.5–1 mm in diameter, disk pale yellowish green, shiny, the thalline rim variable, weakly developed to almost lacking; spores simple, colorless, 4–8 × 8–15 μm, eight per ascus. Thallus K−, C−, P− (usnic acid and undetermined substances).

Habitats: On twigs of oaks and other broadleaf trees and on conifers in North Coastal and Montane Forest and in Valley and Foothill Woodland from 1000 to 6000 ft elevation.

Range: Common from Los Angeles County northward to Lake County in the North and South Coast Ranges and from Kern County to Lassen County in the Sierra Nevada.

Lecidea

Lecidea is a worldwide genus of more than 1000 species. The taxonomy of this heterogeneous group is exceedingly difficult, and few of the species have been given adequate study. The genus can be recognized by the usually dark disk, with a more or less distinct black proper rim, and simple, colorless spores. Externally it is very similar to *Buellia, Rhizocarpon,* and other genera, which have different spores. Several segregate genera are recognized in this guide, in particular the squamulose species *Hypocenomyces, Lecidoma, Psora,* and *Psorula,* as well as crustose *Lecidella, Micarea,* and *Porpidia.* This still leaves about 70 species of *Lecidea* in the state, many of these doubtfully identified by older workers and few recollected by modern workers. The three species included here are by far the commonest ones in California. Additional keys may be found in the old floras of Hasse (1913) and Herre (1910) and the more modern studies by Weber (1963) and Wetmore (1967).

[1. Paraphyses branched and netlike. . . . . *Micarea*]
1. Paraphyses unbranched.
   2. Thallus dark brown, areolate and chinky to subsquamulose. . . . . *L. atrobrunnea*
   2. Thallus whitish or greenish gray or lacking.
      3. Thallus lacking at the rock surface, only black apothecia present.
         4. Ascus with eight spores. . . . . *L. auriculata*
        [4. Ascus with numerous bacilliform spores. . . . .
        . . . . *Polysporina* and *Sarcogyne*]
      3. Thallus distinct, continuous, growing on rocks or trees.
        5. Collected on trees. . . . . *Lecidella elaeochroma*
        5. Collected on rocks.
           6. Thallus thick, chinky; spores without a halo.
           . . . . *Lecidea tessellata*
          [6. Thallus thin, continuous; spores with a thick, gelatinous halo. . . . . *Porpidia*]

Lecidea atrobrunnea (DC.) Schaer.   (fig. 68)

Thallus crustose, chinky to almost squamulose, consisting of crowded, dark brown areoles about 1 mm across with a whitish rim, forming colonies 2–6 cm broad. Apothecia scattered among areoles, angular, 1 mm in diameter, top of hymenium greenish; spores simple, colorless, 3–4 × 10–15 μm, eight per ascus. Cortex and medulla K−, C−, P− (confluentic acid and undetermined substances).

Habitats: On granite, sandstone, lava, and other rocks in North Coastal and Montane Forest and Valley and Foothill Woodland from near sea level to over 7000 ft elevation.

Range: Very common from San Diego County northward to Siskiyou County in the North and Coast Ranges, Cascades,

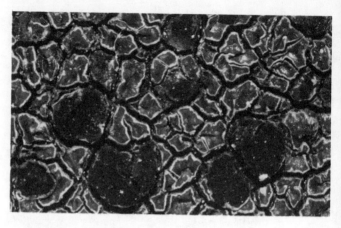

FIG. 68. *Lecidea atrobrunnea* (×10).

and Klamath Mountains and from Kern County to Modoc County in the Sierra Nevada and Modoc Plateau.

Externally this common lichen is very similar to *Rhizocarpon bolanderi*, which has brown, muriform spores.

Lecidea auriculata Th. Fr. (fig. 67f)

Thallus lacking on the rock surface, visible only as a scattered white mat between crystals in the outer layer of rock. Apothecia large, to 1.5 mm in diameter, disk black with a raised black rim; spores simple, colorless, $2-4 \times 6-10$ μm, eight per ascus.

Habitats: On exposed outcrops (granite, sandstone, and other rocks) in North Coastal and Montane Forest, Valley and Foothill Woodland, and Subalpine Forest and Alpine Fell-Field from near sea level to 10,000 ft elevation.

Range: Common in all mountainous areas in the state.

This lichen is recognized by the black apothecia on the surface of rocks and in crevices. The fungal component and algae are scattered just below the surface among crystals of the rock. This growth form is called endolithic. Two externally similar genera, *Sarcogyne* Flot. and *Polysporina* Vězda, have numerous small spores in each ascus; they are widespread in California but not as common as *L. auriculata*. To be safe you should examine all endolithic specimens for spores with a microscope.

Lecidea tessellata Flk. (fig. 69a)

Thallus crustose, thick, deeply fissured and chinky at the center, grayish white, 3–10 cm broad. Apothecia black, conspicuous, 1–2 mm in diameter, round to angular and often produced in concentric rings; spores simple, colorless, 4 × 6–9 μm. Thallus K−, C−, P− (confluentic acid and undetermined substances).

Habitats: On granite and other acidic rocks in rather sheltered outcrops in Valley and Foothill Woodland and in Montane Forest from 1000 to 8000 ft elevation.

Range: Common from Riverside County to Glenn County in the North and South Coast Ranges and Fresno County to Plumas County on the western slopes of the Sierra Nevada.

Lecidella Krb.

This genus was lumped with *Lecidea* until recently. It is differentiated by a more complex chemistry (xanthones present), noncoherent paraphyses, and a noncarbonized exciple. Externally, it closely resembles *Lecidea*. Several species are known from California, of which corticolous *L. elaeochroma* is by far the most common. The few species on rock may be confused with *Lecidea auriculata*.

Lecidella elaeochroma (Ach.) Choisy (fig. 69b)

Thallus minutely warty, greenish gray, 1–3 cm broad. Apothecia common, to 1 mm in diameter, disk black, flat to convex, the black proper rim barely visible; epithecium bluish green, spores colorless, simple, 6–10 × 9–15 μm, eight per ascus. Thallus K−, C−, P− (substances not determined).

Habitats: On oaks and other broadleaf trees and on conifers in North Coastal and Montane Forest and in Valley and Foothill Woodland from sea level to 6000 ft elevation.

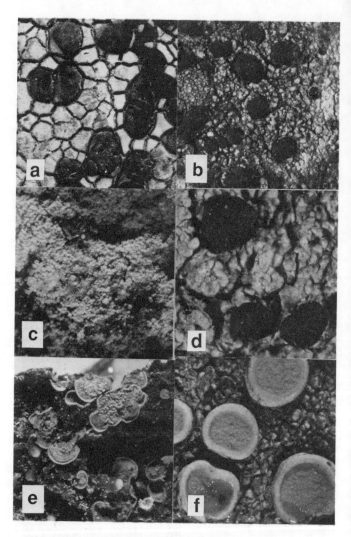

FIG. 69. Species of crustose lichens: (a) *Lecidea tessellata;* (b) *Lecidella elaeochroma;* (c) *Lepraria membranacea;* (d) *Mycoblastus sanguinarius;* (e) *Normandina pulchella;* (f) *Ochrolechia pallescens* (all ×10).

Range: Common from Ventura County northward to Mendocino County in the North and South Coast Ranges and from Tulare County to Plumas County on the western slopes of the Sierra Nevada.

Lepraria Ach.

Lepraria is an old generic name for the so-called imperfect crusts which have never been found in fruit. The thallus consists of a whitish gray or yellow, granular-sorediate cover on soil, rocks, or the base of trees. Surprisingly, the chemical variation in this presumably primitive group is very great, suggesting the existence of many populations. A bright deep lemon yellow population is called *L. chlorina* (Ach.) Ach.

Lepraria membranacea (Dicks.) Vain.   (fig. 69c)

Thallus crustose, powdery, granular-sorediate, pale greenish to grayish white, 2–10 cm broad. Fruiting bodies unknown. Thallus K+ yellow or K−, C−, P+ orange or P− (chemistry not determined).

Habitats: Base of trees, humus and soil (often on overhanging banks), and on sheltered rocks from sea level to 10,000 ft.

Range: Found throughout the state.

Mycoblastus Norm.

This small crustose genus is closely related to *Lecidea*. The large black apothecia are conspicuous, but the most important diagnostic character is the large, thick-walled spores. There are two species in California, rather common in the North Coastal Forest, one with a red layer under the hymenium (*M. sanguinarius*) and one lacking it (*M. alpinus* (Fr.) Kernst.).

1. Layer under the hymenium bright red. . . . .
. . . . *M. sanguinarius*
[1. Layer under the hymenium brownish. . . . . . *M. alpinus*]

Mycoblastus sanguinarius (L.) Norm.   (fig. 69d)

Thallus whitish gray, continuous, thickish and verrucose, 6–10 cm broad. Apothecia conspicuous, black, 1–2 mm in diameter, strongly convex with the proper rim disappearing; lower part of hymenium bright red; spores simple, colorless, 20 × 50–60 μm, two per ascus. Medulla K−, C−, P− (caperatic acid).

Habitats: On Douglas Fir and other conifers in North Coastal Forest from near sea level to 1000 ft elevation.

Range: Rather common in Del Norte and Humboldt counties in the North Coast Ranges.

Normandina Nyl.

This curious lichen is often not seen in the field but discovered later in the laboratory mixed with other lichens. The minute, delicate squamules with an upturned rim are unmistakable. *Normandina* has apparently never been collected with fruiting bodies but is usually classified with the basidiomycete fungi in the order Agaricales.

Normandina pulchella (Borr.) Nyl. (fig. 69e)

Thallus consisting of tiny, fragile, scattered squamules attached at the center below, about 1 mm in diameter with a delicate upturned rim; soredia developing at the center of the squamules. Thallus K−, C−, P− (no substances present).

Habitats: On mossy trunks of oak and other broadleaf trees in North Coastal and Montane Forest and Valley and Foothill Woodland from near sea level to 3000 ft elevation.

Range: Widespread but often overlooked from the San Francisco Bay area to Mendocino and Del Norte counties in the North Coast Ranges and Plumas County in the Sierra Nevada.

Ochrolechia Mass.

This is one of the more conspicuous crustose genera on trees in California. The apothecia are very large (2–3 mm in diameter) and the disk whitish pink. The thalline rim, pale greenish gray, is well developed. The spores are large and thick-walled. The only similar species, *Lecanora caesiorubella,* has much smaller spores. Howard (1970) listed eight species for California, but only one, *O. pallescens,* is very common. The other species are much less frequently encountered in the field and in fact need more study.

Ochrolechia pallescens (L.) Mass. (fig. 69f)

Thallus whitish gray, 2–6 cm broad; surface smooth and shiny, cracked with age. Apothecia common, 1–2 mm in diameter with a thick thalline rim, disk pinkish to white pruinose; spores 15–30 × 50–80 μm. Disk and thallus K−, C+, KC+ red, P− (gyrophoric acid).

Habitats: On oaks and other broadleaf trees and on conifers in North Coastal and Montane Forest from near sea level to 5000 ft elevation.

Range: Rather common from San Diego County northward to Del Norte County in the North and South Coast Ranges and in Butte and Plumas counties on the western slopes of the Sierra Nevada.

Pachyospora Mass.

This genus contains just one species, which grows on detritus and rotted bark of broadleaf trees. The flush to sunken black apothecia closely resemble those of *Aspicilia,* a saxicolous genus, except that with age they become white pruinose. The spores are much larger than in *Aspicilia.*

Pachyospora verrucosa (Ach.) Mass. (fig. 70a)

Thallus whitish gray, continuous to lobulate and warty, 3–8 cm broad. Apothecia flush to sunken, 0.5–1 mm in diameter, the thalline rim broad, often white pruinose, disk convex, black, dull; spores colorless, simple, 18–25 × 45–50 μm. Thallus K−, C− (no substances present).

Habitats: On rotted bark of broadleaf trees in Valley and Foothill Woodland from 1000 to 5000 ft elevation.

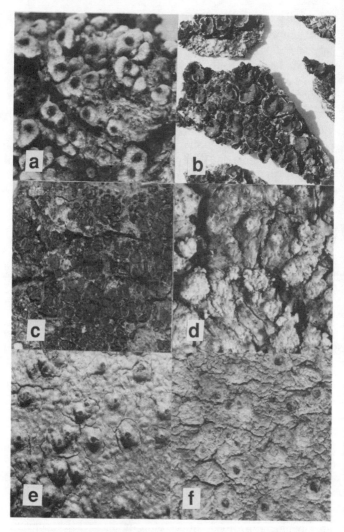

FIG. 70. Species of crustose lichens: (a) *Pachyospora verrucosa*; (b) *Peltula euploca*; (c) *P. polyspora*; (d) *Pertusaria amara*; (e) *P. pustulata*; (f) *Phlyctis argena* (all ×10).

Range: Rather rare but probably overlooked from Lake County into the Klamath Mountains and Cascades and from Calaveras County to Shasta County on the western slopes of the Sierra Nevada.

Peltula Nyl.

Peltula is a small genus of soil-inhabiting species, most of them squamulose but one (*P. euploca*) clearly umbilicate. It is placed with *Heppia* in the family Heppiaceae. The algal symbiont is blue-green, and the peculiar apothecia are immersed, flush, or at most sessile on the squamules. There are 8 spores per ascus in *Heppia* and 16 or more in *Peltula;* spore number is an important difference between the two genera. Eight species known in California have been studied in detail by Wetmore (1970), and this revision should be consulted for more details.

Superficially *Heppia* and *Peltula* are close to *Catapyrenium* (black perithecia present) and less so to *Psora* (apothecia present), and all occur together on soil in semiarid and arid parts of California. They have considerable ecological importance in consolidating soil but are quickly destroyed by dune buggies and motorcycles.

1. Thallus umbilicate, the margins sorediate; collected on rocks. . . . . *P. euploca*
1. Thallus squamulose, adnate; collected on soil.
   [2. Asci with eight spores. . . . . *Heppia*]
     2. Asci with numerous spores. . . . . *P. polyspora*

Peltula euploca (Ach.) Poelt   (fig. 70b)

Thallus dark brown, umbilicate at maturity, to 1 cm broad, the surface smooth to slightly papillose, margins with powdery gray soredia; lower surface light brown, smooth. Apothecia lacking. Algal component blue-green. Cortex and medulla K−, C−, P− (chemistry unknown).

Habitats: On rocks in Valley and Foothill Woodland from 1000 to 3000 ft elevation.

Range: Rather rare but probably overlooked from San Diego County to the Santa Cruz Mountains in the North and South

Coast Ranges and in Tulare County in the Sierra Nevada foothills.

Peltula polyspora (Herre) Wetm. (fig. 70c)

Thallus growing on and consolidating soil, dark brown to blackish, consisting of squamules 1–2 mm wide forming colonies 2–6 cm across, curling up somewhat at maturity. Apothecia produced in the middle of squamules, sunken to flush, about 0.5 mm wide, disk light brown; spores simple, colorless, globose, 6 μm long, about 100 per ascus. Thallus K−, C−, P− (no substances present).

Habitats: On soil in semiarid and desert localities near sea level.

Range: Widespread from San Diego and Imperial counties northward to Tehama and Sierra counties in the North and South Coast Ranges and the western slopes of the Sierra Nevada.

Pertusaria DC.

This widespread crustose genus can be recognized by the warty protuberances on the thallus, which contain immersed apothecia or are sorediate. The spores are large and thick-walled, often only one or two per ascus. There are at least 12 species in California according to a recently published monograph (Dibben, 1980), which should be consulted by serious students. The species are differentiated by a complex series of apothecial, spore, and chemical characters and occur most commonly in the South Coast Ranges from Marin County southward.

1. Apothecia disk-shaped or forming sorediate warts (hand lens not needed).
   2. Warts coarsely sorediate; collected on trees. . . . .
   . . . . *P. amara*
   [2. Warts not sorediate; collected on rocks. . . . .
   . . . . *P. flavicunda*]
1. Apothecia enclosed in warts, opening through pores; soredia lacking.

3. Collected on tree bark. . . . . *P. pustulata*
[3. Collected on rocks. . . . . *P. californica*]

Pertusaria amara (Ach.) Nyl. (fig. 70d)

Thallus greenish gray, 4–8 cm wide; surface shiny, continuous to cracked and finely white-spotted, densely sorediate, the soralia 1–2 mm wide. Apothecia not seen. Medulla K−, C−, KC+ red, P− (picrolichenic acid).

Habitats: On broadleaf trees and conifers in Valley and Foothill Woodland and North Coastal Forest from near sea level to 2500 ft elevation.

Range: Rather common from Santa Barbara County northward to Yuba and Humboldt counties in the North and South Coast Ranges and from Plumas County to Shasta County in the western foothills of the Sierra Nevada.

A very similar sorediate species, *P. ophthalmiza* (Nyl.) Nyl., lacks any chemistry.

Pertusaria pustulata (Ach.) Duby (fig. 70e)

Thallus crustose, continuous, shiny, pale tannish gray, 2–6 cm broad. Apothecia immersed in large warts, 1–1.5 mm in diameter, two to four apothecia per wart, each opening by a tiny dark pore; spores colorless, simple, large, $35 \times 100–150 \ \mu m$, two per ascus. Thallus K+ yellow to red, C−, P+ orange (stictic acid).

Habitats: On oaks and other broadleaf trees in Valley and Foothill Woodland from near sea level to 1000 ft elevation.

Range: Rather common from Los Angeles County north to the San Francisco Bay area in the South Coast Ranges.

Phlyctis (Wallr.) Flot.

This crustose genus, generally considered to be a member of the family Lecanoraceae, has inconspicuous, flush apothecia. The disk becomes granular white pruinose with age and may not even be recognized as being part of the apothecia. The large, colorless muriform spores are the most characteristic feature. A second, rarer species, *P. agelaea* (Ach.) Flot., has spores less than 100 μm long.

Phlyctis argena (Spreng.) Flot. (fig. 70f)

Thallus crustose, grayish white, finely verruculose and cracked, 2–8 cm broad. Apothecia numerous, flush to slightly emergent, about 0.5 mm in diameter, the disk dark or becoming white pruinose, the rim granular and sorediate, sometimes hiding the disk below; spores colorless, densely muriform, 30 × 150 μm, one per ascus. Thallus K+ yellow turning red, C−, P+ orange (norstictic acid).

Habitats: On oaks, California Laurel, and other broadleaf trees in North Coastal Forest, Valley and Foothill Woodland, and Coastal Scrub from near sea level to 3000 ft elevation.

Range: Widespread in Los Angeles and Ventura counties, the Channel Islands, and the San Francisco Bay area northward to Humboldt County in the North and South Coast Ranges.

Placopsis (Nyl.) Linds.

This is a marginally lobate, crustose lichen which would probably be considered foliose at first glance. There is no lower cortex, however, and the thallus is very difficult to remove from the rock. The most obvious distinguishing character is the large, brown, warty cephalodia.

Placopsis gelida (L.) Linds. (fig. 71a)

Thallus light greenish brown, tightly adnate on rock, 1–3 cm broad; marginal lobes 1–1.5 mm wide, splayed at the tips, a lower cortex lacking; center of thallus areolate-cracked, the surface with capitate soralia to 1 mm broad and large brown warty cephalodia. Apothecia lacking. Medulla K−, C+ rose, P− (gyrophoric acid).

Habitats: On moist rocks in North Coastal Forest near sea level.

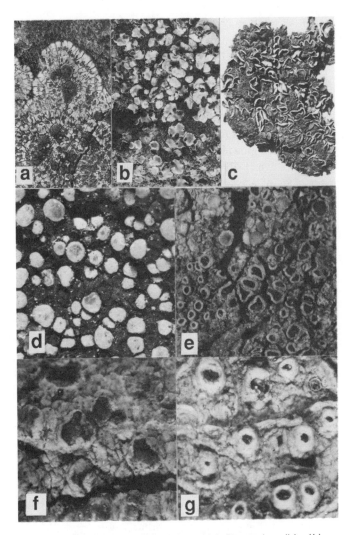

FIG. 71. Species of crustose lichens: (a) *Placopsis gelida;* (b)
Psora decipiens; (c) *P. nipponica;* (d) *Rhizocarpon disporum;*
(e) *Rinodina oregona;* (f) *Roccellina franciscana;* (g) *Thelotrema lepadinum* (a–c: ×1; e–g: ×100).

Range: Widespread from San Mateo County to Del Norte County in the North Coast Ranges.

Psora Hoffm.

Psora is allied to and sometimes included in the genus *Lecidea*. It has typical *Lecidea*-like apothecia with simple, colorless spores, but the thallus is distinctly squamulose. The squamules are corticate below and vary from adnate to suberect, almost always growing on soil or in rock crevices. Two recent segregates, which also occur in California, lack a lower cortex: *Lecidoma*, which has flush laminal apothecia, and *Psorula*, which has marginal apothecia. As with many other crustose genera, the taxonomy of this group of soil-inhabiting lichens is very poorly known. They do, however, occur widely, especially in drier areas, where they act as soil consolidators. In the same habitats one may find *Catapyrenium, Heppia,* and *Peltula,* all very similar in appearance.

1. Squamules suberect, curling up with a conspicuous whitish lower surface; medulla C+ rose . . . . *P. nipponica*
1. Squamules adnate to barely suberect, not curling up; mostly dark below but sometimes with a whitish rim; medulla C−.
   2. Squamules pink to reddish above or becoming densely white pruinose. . . . . *P. decipiens*
   2. Squamules reddish brown to blackening; pruina usually not well developed.
      3. Lower cortex lacking.
         [4. Apothecia flush, laminal. . . . .
         . . . . *Lecidoma demissa*]
         [4. Apothecia marginal. . . . . . *Psorula rufonigra*]
      3. Lower cortex present.
         [5. Squamules white below. . . . . *Psora russellii*]
         [5. Squamules reddish brown below. . . . .
         . . . . *P. luridella*]

Psora decipiens (Hedw.) Hoffm. (fig. 71b)

Squamules brownish to pink or reddish, 1–7 mm wide, closely adnate on soil forming colonies 2–6 cm broad, crowded or separate, the surface often becoming densely white pruinose;

lower surface tan. Apothecia common, marginal on the squamules, to 1 mm in diameter, disk dark; spores colorless, simple, 6 × 12 μm. Medulla K+, P+ yellow, C− (norstictic acid).

Habitats; On soil in semiarid and desert regions from near sea level to 10,000 ft elevation.

Range: Widespread from San Diego County north to Inyo and Kern counties.

Psora nipponica (Zahlbr.) Schneid.   (fig. 71c)

Squamules brown, adnate to suberect with upturned, curling margins, 3–8 mm wide, forming colonies 2–6 cm broad, the lower surface white. Apothecia common, to 1.5 mm in diameter, disk black; spores colorless, simple, 6 × 12 μm.

Habitats: On soil over rocks and in rock crevices in Montane Forest at 3000 to 7000 ft elevation.

Range: Widespread in Lake County in the North Coastal Ranges and from Mariposa County to Shasta County on the western slopes of the Sierra Nevada.

Rhizocarpon Ram.

Rhizocarpon is primarily an arctic genus but also occurs widely in the mountains of western North America. It is characterized by a well-developed black hypothallus over which the bullate-areolate, chinky thallus is scattered. The black apothecia are very similar externally to those of *Lecidea,* but the spores are muriform and often brown and the paraphyses branched. *Rhizocarpon geographicum* is one of the commonest and most conspicuous lichens in the High Sierra, forming deep yellow colonies on exposed rocks. The chinky yellow areoles are separated by the black lines of the hypothallus, suggesting a map and giving rise to the common name Map Lichen. There are also several gray species in California, but as with most other crustose genera, the lack of any modern taxonomic treatment makes accurate species identification very difficult.

1. Thallus deep greenish yellow to yellow. . . . .
. . . . *R. geographicum*
1. Thallus mineral or brownish gray.
    2. Asci containing two spores. . . . . *R. disporum*
    2. Asci containing four to eight spores.
        [3. Thallus consisting of small, flattened, brown, white-
        rimmed areoles; medulla K+ yellow. . . . .
        . . . . *R. bolanderi*]
        [3. Thallus consisting of gray bullate areoles; medulla
        K−. . . . . *R. grande*]

Rhizocarpon disporum (Hepp) Müll. Arg.   (fig. 71d)

Thallus crustose, consisting of small gray bullate areoles or warts scattered on a black hypothallus, 2–8 cm broad. Apothecia produced between warts, about 0.5 mm in diameter, disk black, the proper rim scarcely noticeable, black; spores brown, muriform, 15–20 × 50–60 μm, two per ascus. Medulla K+ yellow, C−, P+ orange (stictic and norstictic acids).

Habitats: On granite and other acidic rocks in open sunny places in Valley and Foothill Woodland and Montane Forest from near sea level to 7000 ft elevation.

Range: Rather common from San Diego County northward to the San Francisco Bay area and from Inyo County to Modoc County in the Sierra Nevada and Modoc Plateau.

Another common species in the Sierras, *R. grande* (Flot.) Arn., reacts C+ red (gyrophoric acid) in the medulla. A widespread species with white-rimmed areoles as in *Lecidea atrobrunnea*, *R. bolanderi* (Tuck.) Herre, reacts K+ yellow (stictic acid). These two species must be examined for spores for positive identification.

Rhizocarpon geographicum (L.) DC.   (pl. 9c)

Thallus crustose, chinky, consisting of small yellow areoles scattered on a black hypothallus 2–10 cm broad. Apothecia occurring between areoles, 0.5–1 mm in diameter, disk black, lacking a distinct rim; spores brown, muriform, 10 × 25–35 μm, four per ascus. Thallus K−, C−, P− (rhizocarpic acid).

Habitats: On exposed granite and other acidic rocks in Montane Forest, Subalpine Forest, and Alpine Fell-Field from 1000 to 11,000 ft elevation.

Range: Common from Riverside County northward to Trinity County in the North and South Coast Ranges and Trinity Mountains and from Fresno and Inyo counties to Modoc County in the Sierra Nevada and Modoc Plateau.

Rinodina (Ach.) S. Gray

This crustose genus is closely related to *Buellia* but differs in having a thalline rim around the black disk. The thallus is usually a dark greenish or brownish gray, continuous to chinky, fissured, or even areolate. Except for the black disk and dark greenish thallus, it could be mistaken for a *Lecanora.* However, the spores are brown and two-celled. Specimens are often overlooked in the field, only to be discovered later in the laboratory as mixtures with *Buellia, Lecanora,* and other crusts on twigs and bark.

There are at least 10 species in California, but we lack a modern study of their taxonomy. The commonest species appears to be *R. oregona.* Another common corticolous species, *R. hallii* Tuck., has an inconspicuous thalline rim. *Rinodina marysvillensis* Magn. is unique in having a white thallus and a white rim around the large black apothecia. Several species grow on soil (*R. bolanderi* Magn. and *R. conradi* Kbr.) or rocks (*R. bischoffii* (Hepp) Mass.).

Rinodina oregona **Magn.**   (fig. 71e)

Thallus brownish-greenish gray, 2–3 cm broad; surface continuous, becoming areolate, dull. Apothecia numerous with a distinct thalline rim, about 0.5 mm in diameter, disk plane, black; spores brown and two-celled, $10-12 \times 18-26$ μm. Thallus K−, C−, P− (no substances present).

Habitats: On oaks and other broadleaf trees and conifers in Montane Forest and Valley and Foothill Woodland from near sea level to 3000 ft elevation.

Range: Widespread from San Bernardino County to the San Francisco Bay area in the South Coast Ranges and in Plumas and Shasta counties in the Sierra Nevada.

Roccellina Darb.

This small genus is representative of the crustose members of the family Roccellaceae. As in fruticose *Roccella,* the apothecia are round to irregular, sessile, and have a stroma-like, branched network of pseudoparaphyses in the hymenium. This character has to be determined with a microscope. *Roccellina* and a very closely related genus, *Dirina* Fr., were recently revised by Tehler (1983). Both are rather common on trees and rocks from the Monterey area south to Catalina Island. *Lecanactis,* another member of this family, has a dark disk and raised proper rim but no thalline rim. It may be misidentified as a *Bacidia* if the paraphysoid structure is not correctly observed.

Roccellina franciscana (Herre) Huneck & Follm. (fig. 71f)

Thallus crustose, rather thick, bullate and chinky, grayish white, 5–10 cm broad. Apothecia sessile, round to irregularly lobed, large, to 2 mm across with a distinct thalline rim, disk brownish to white pruinose; asci scattered in a dense stroma of branched, adhering pseudoparaphyses; spores colorless, three-septate, 6–8 × 20–30 μm, eight per ascus. Thallus K+ yellowish, P−, C (substances not determined).

Habitats: On conifers in forests near sea level.

Range: Common in Monterey County in the South Coast Ranges.

Thelotrema Ach.

This crustose genus can be recognized in the field with a hand lens: The sunken apothecia resemble small volcanoes. A single species of this otherwise large tropical family, *T. lepadinum,* occurs in temperate areas.

Thelotrema lepadinum (Ach.) Ach. (fig. 71g)

Thallus crustose, very thin, pale tannish gray, 6–10 cm broad; apothecia urn-shaped at maturity, 0.5–1 mm in diameter, opening with a broad pore to give the appearance of a volcano, disk recessed with the thin proper exciple appearing as an inner ring; spores colorless, muriform, $15-30 \times 60-80 \ \mu$m. Thallus K−, C−, P− (no substances present).

Habitats: On bark of hardwood trees (maple, *Myrica*, and others) in the North Coastal Forest from sea level to 300 ft elevation.

Range: Widespread from San Mateo County northward to Humboldt County in the North Coast Ranges.

Glossary

Adnate: lying flat on and attached to the substratum.

Apical: at the tip or terminal part of a lobe or podetium.

Apothecium: the reproductive fruiting body of Ascomycete fungi and lichens, usually disk or cup-shaped (pl. apothecia).

Appressed: lying flat on and firmly attached to the substratum.

Areolate: composed of areoles, as the surface of some Cladonias and the thallus of some crustose lichens.

Areole: individual segmented parts of the thallus formed by cracks.

Articulated: composed of segments, as branches of *Usnea*.

Ascolocular: lichens having a hymenium with asci produced in cavities among branched pseudoparaphyses.

Ascospore: a spore produced in an ascus through meiotic stages.

Ascus: a sac containing spores, located in the hymenium (pl. asci).

Axil: the angle between branches (in *Cladonia* and *Cladina*) or lobes (in foliose lichens).

Bullate: inflated or blisterlike, as areoles.

Calcareous: referring to rocks or soil containing calcium minerals.

Canaliculate: grooved on the lower surface, as *Evernia*.

Capitate: shaped like a head, usually referring to soralia.

Carbonaceous: blackened and usually hard.

Cephalodium: tiny thalli containing blue-green algae growing on or in the thallus of *Lobaria, Peltigera, Placopsis,* and *Stereocaulon* (pl. cephalodia).

Chinky: cracked and fissured, as the center of the thallus in some crustose lichens.

Cilia: hairlike outgrowths along the margins of lobes.

Ciliate: provided with cilia.

Clavate: club-shaped, with a tapering base and enlarged tip.

Coalesce: to fuse together, as many small thalli merging into a large colony.

Colony: a single lichen thallus or several growing closely together.

Conidia: unicellular bacilliform cells produced in pycnidia.

Contiguous: touching or in close contact, as lobes.

Convex: curving outward, as the disk of some apothecia.

Convoluted: rolled up lengthwise or twisted.

Coralloid: resembling coral, as richly branched isidia.

Cord: a dense strand of hyphae in the center of branches of *Usnea*.

Cortex: outermost layer of the thallus, consisting of compressed hyphal cells.

Corticate: having a distinct cortex.

Corticolous: growing on tree bark.

Crenulate: having a notched margin, as lobes.

Crustose: a lichen growth form, the thalli growing in intimate contact with the substratum and lacking a lower cortex.

Cyphella: large, circular, recessed pore in the lower surface of *Sticta* (pl. cyphellae).

Dichotomous: dividing into two parts, as the branching pattern in lobes, branches, or rhizines of lichens.

Diffuse: scattered without a definite pattern, as diffuse soredia.

Disk: the round surface of an apothecium.

Dissected: cut up or divided into many small lobes.

Dorsiventral: having distinct upper and lower surfaces that are different in appearance.

Eciliate: lacking cilia, usually describing lobe margins.

Ellipsoidal: shaped like an ellipse, referring to spores.

Endemic: occurring only in a small geographic area.

Endolithic: growing just under the rock surface, referring to a group of crustose lichens.

Entire: continuous, without breaks.

Epithecium: layer of tissue on surface of the hymenium (disk) of an apothecium.

Exciple: a margin around the apothecial disk.

Excrescence: an amorphous outgrowth, as the moldlike material in *Niebla*.

Farinose: flourlike, referring to consistency of soredia.

Fibril: a short, isidia-like, lateral branch, as in *Usnea*.

Fibrillose: having fibrils.

Fissured: deeply cracked, as in the cortex of various lichens (see also *areolate*).

Foliose: a leaflike lichen growth form, usually growing adnate on the substratum.

Foveolate: finely pitted or dimpled.

Fruiting body: spore-producing bodies (apothecia and perithecia).

Fruticose: a shrubby or hairlike lichen growth form, attached only at the base or free growing and pendulous to erect.

Furcate: regularly forked, referring to branching patterns of lobes and podetia.

Gelatinous: jellylike, used to describe unstratified lichens containing blue-green algae when wet.

Granular: grainy, used to describe coarse soredia.

Hairs: fine cellular outgrowths from the cortex, as in *Phaeophyscia*.
Hapter: a whitish tuft of hyphae on the lower surface in *Collema* and *Leptogium* (pl. haptera).
Hymenium: the fertile layer in the apothecium (and perithecium) consisting of asci and paraphyses.
Hypha: a microscopic multicellular fungal thread making up the lichen thallus (pl. hyphae).

Immersed: sunken in the thallus, as fruiting bodies.
Isidiate: having isidia.
Isidioid: resembling isidia.
Isidium: a cylindrical, fingerlike outgrowth from the upper cortex 1–2 mm high (pl. isidia).

Labriform: lip-shaped.
Lacerate: with jagged edges or tips, as in lobe margins and podetia.
Laciniate: divided into numerous small lobes.
Laminal: superficial on the surface of the thallus (as opposed to marginal).
Lateral: produced along the margins.
Lecideine: having only a proper margin or exciple, referring to apothecia.
Linear: narrow and having a uniform width, as lobes or soralia.
Lirella: elongate fruiting body of *Graphis* (pl. lirellae).
Lirelliform: shaped like lirellae.
Lobate: forming lobes, usually referring to the margin of crustose lichens.
Lobe: a rounded or strap-shaped division of the foliose thallus.
Lobulate: having small lobes produced on the surface or margin of main lobes.
Locule: individual cells in ascospores.

Maculate: white-spotted, as in the surface of *Physcia* species.
Marginal: located along lobe margins, as soralia.
Mazaedium: a loose mass of spores on the apothecial disk of the order Caliciales (pl. mazaedia).
Medulla: inner portion of the thallus consisting of loosely interwoven hyphae.
Mottled: variegated white and black or brown, as the lower surface of some foliose lichens.
Muriform: divided into many chambers, as spores.

Nonisidiate: lacking isidia.
Nonsorediate: lacking soredia.

Orbicular: circular in outline, as the shape of a thallus or soralium.

Palisade cortex: a cortex with cells arranged in vertical tiers.

Papilla: a small rounded bump on the cortex (pl. papillae).

Papillose: having papillae.

Paraphysis: threadlike hypha packing spaces between asci in the hymenium of apothecia (pl. paraphyses).

Paraplectenchymatous: cortical structure with irregularly arranged, compressed cells.

Pendulous: hanging from or draping tree trunks or rocks.

Perforate: pierced with holes.

Perithecium: flask-shaped fruiting body of pyrenocarp lichens (pl. perithecia).

Phycobiont: the algal partner in lichens located in the algal layer.

Phyllocladia: tiny granular or leaflike structures on branches of *Stereocaulon*.

Plates: flattened rhizine-like structures on the lower surface of some Umbilicarias.

Podetium: hollow, simple or branched, upright structure in *Cladonia* arising from the squamules (pl. podetia).

Polarilocular: two-celled spores with a central perforated septum (wall).

Primary thallus: squamules or granular thallus of *Baeomyces, Cladonia, Pilophoron,* and *Stereocaulon.*

Prosoplectenchymatous: cortical structure with cells in a parallel arrangement.

Prostrate: lying flat on the substratum.

Pruina: fine white frosty or granular covering on upper cortex or apothecia, especially in *Physcia.*

Pruinose: having pruina.

Propagule: isidia, soredia, or thallus fragments which propagate lichens.

Proper rim: the alga-free margin of apothecia (see also *exciple*).

Pseudoapothecia: apothecia-like structures produced in *Roccella* and other lichens.

Pseudocyphella: small white pore in the upper or lower cortex (pl. pseudocyphellae).

Pseudocyphellate: having pseudocyphellae.

Pseudoparaphysis: branched, paraphysis-like thread in the hymenium of ascolocular lichens (pl. pseudoparaphyses).

Pseudopodetium: upright, fruticose thallus of *Stereocaulon* (pl. pseudopodetia).

Pustulate: covered with blisterlike structures.

Pycnidium: small, flask-shaped reproductive structures in the medulla (or on margins of lobes in *Tuckermannopsis*) containing conidia (pl. pycnidia).

Pyrenocarp: a lichen with perithecia as the fruiting body (adj.: pyrenocarpous).

Reticulate: arranged in a network.

Revolute: rolled downward, as lobe tips.

Rhizine: compact strands of hyphae on the lower surface of many foliose lichens; an attachment organ.

Rim: the margin or exciple of an apothecium.

Rotund: rounded in outline, as tips of broad-lobed lichens.

Rugose: wrinkled or ridged.

Saxicolous: growing on rocks.

Scabrid: having fine white scales on the upper cortex, as in *Peltigera* and *Physconia*.

Septate: divided into two or more locules by a wall, as spores. A spore with two walls is referred to as two-septate, one with three walls as three-septate, and so forth.

Sessile: located on the thallus surface, as apothecia (as opposed to stalked or immersed).

Simple: unbranched or in spores having only one locule.

Soralium: powdery clumps of soredia on the surface or margins of thalli (pl. soralia).

Sorediate: having soredia.

Soredium: a microscopic clump of several algal cells surrounded by hyphae, erupting at the thallus surface as a powder to form soralia (pl. soredia).

Spatulate: shaped like a spatula.

Spinulate: having spinules.

Spinule: short, isidia-like branch, as in *Usnea* (see also *fibril*).

Splayed: having tips expanded, as in rhizines of *Peltigera*.

Spore: microscopic reproductive cell of a fungus produced in an ascus.

Squamule: a scalelike thallus structure in lichens, as *Cladonia* and *Psora*.

Squamulose: provided with squamules; also referring to growth form.

Squarrose: branching pattern of rhizines where small branches arise perpendicular to the main rhizine.

Striate: having striations.

Striation: an elongate, white ridge on the cortex, as in *Ramalina*.

Stroma: a mass of sterile hyphae where fruiting bodies are produced.

Subascending: with lobe tips raised somewhat above the substratum.

Subcrustose: referring to crustose lichens having a lobate margin.

Suberect: ascending strongly toward the thallus margin, as in *Parmotrema*.

Subfoliose: referring to crustose lichens with a lobate margin.

Subfruticose: intermediate growth form between foliose and fruticose.

Subsorediate: imperfectly sorediate.

Substrate: the medium (soil, rock, bark) on which a lichen grows.

Symbiosis: an association of two different organisms living together in harmony for mutual benefit.

Terminal: at the end of a branch or podetium.

Thalline rim: rim or margin of an apothecium containing thallus tissue (fungus and alga), as opposed to the alga-free lecideine or proper rim.

Thallus: the plant body of a lichen, usually classified as foliose, fruticose, or crustose (pl. thalli).

Tomentose: provided with tomentum.

Tomentum: multicellular, woolly or felty hairs on the lower surface of *Lobaria, Nephroma, Pseudocyphellaria,* and *Sticta* and on the pseudopodetia of *Stereocaulon.*

Transversely septate: spores with ladderlike cross walls.

Umbilicate: a growth form where the thallus is attached to the substratum by a central umbilicus.

Umbilicus: a single strand of rhizines on the lower surface of umbilicate lichens.

Vein: raised, riblike structure on the lower surface of *Peltigera.*

Verrucose: covered with wartlike outgrowths.

Verruculose: covered with fine warts.

White-spotted: having numerous tiny white spots on the upper cortex, as in *Physcia* (see also *maculate*).

Suggested References

This list includes general references to lichens as well as those dealing more specifically with the California or western U.S. species. Unfortunately, many of the books are out of print and the journal articles are often available only in university libraries. Moreover, the older works such as Hasse (1913) and Herre (1910) use outdated names and a much broader species concept than we use today.

Ahmadjian, V. and M. E. Hale. 1974. The lichens. Academic Press, New York. [A college-level text on most aspects of lichenology.]

Bolton, E. M. 1960. Lichens for vegetable dyeing. Charles T. Branford Co., Newton Centre, Mass. [Practical guide for lichen dyeing.]

Brodo, I. M. 1984. The North American species of the *Lecanora subfusca* group. *Beihefte Nova Hedwigia* 79:63–185.

———— and D. Hawksworth. 1977. *Alectoria* and allied genera in North America. *Opera Botanica* 42 (Lund, Sweden). [Complete monograph of *Alectoria, Bryoria, Pseudephebe,* and *Sulcaria*.]

Culberson, C. F. 1969. Chemical and botanical guide to lichen products. University of North Carolina Press, Chapel Hill. [Includes an index by species with lichen substances.]

Culberson, W. L. and C. F. Culberson. 1968. The lichen genera *Cetrelia* and *Platismatia*. *Contrib. U.S. Nat. Herb.* 34:449–558. [Illustrated monograph including the California species.]

Degelius, G. 1974. The lichen genus *Collema* with special reference to the extra-European species. *Symbol. Bot. Upsal.* 20(2):1–215. [A world monograph with good keys and notes on California species.]

Dibben, M. 1980. The chemosystematics of the lichen genus *Pertusaria* in North America north of Mexico. *Milwaukee Publ. Mus., Publ. in Biol. & Geol.* 5:1–162. [Monograph with full treatment of California species.]

Esslinger, T. L. 1977. A chemosystematic revision of the brown *Parmeliae*. *Journ. Hattori Bot. Lab.* 42:1–211. [Monograph with detailed information on *Melanelia* and *Neofuscelia*.]

————. 1978. Studies in the lichen family Physciaceae. II. The genus *Phaeophyscia* in North America. *Mycotaxon* 7:283–320.

Fink, B. 1935. The lichen flora of the United States. University of Michigan Press, Ann Arbor. [The only lichen flora of North America, now both out of date and out of print.]

Hale, M. E. 1961. Lichen handbook. Smithsonian Institution, Washington, D.C. (Available from University Microfilms, Ann Arbor.)

[Guide intended for beginners with keys to lichens of eastern North America.]

————. 1979. How to know the lichens. W. C. Brown Company, Dubuque, Iowa. [Pictured keys to the lichens of North America, but not including crusts.]

————. 1983. The biology of lichens. Ed. Arnold (Publishers), 300 North Charles Street, Baltimore, MD 21201. [College-level introduction to lichenology.]

Hasse, H. E. 1913. The lichen flora of southern California. *Contr. U.S. Nat. Herb.* 17:1–132. [Historical treatment with keys and descriptions (but no illustrations), centered on Los Angeles and San Bernardino counties.]

Herre, A.W.C.T. 1910. The lichen flora of the Santa Cruz peninsula, California. *Proc. Washington Acad. Sci.* 12(2):27–269. [Historical treatment with keys and descriptions, but no illustrations.]

Howard, G. E. 1950. Lichens of the state of Washington. University of Washington Press, Seattle. [Partially illustrated flora with keys to all lichen groups.]

————. 1970. The lichen genus *Ochrolechia* in North America north of Mexico. *Bryologist* 73:93–130. [Monograph of the genus with full coverage of California species.]

Imshaug, H. A. 1951. The lichen-forming species of the genus *Buellia* occurring in the United States and Canada. University Microfilms, Ann Arbor.

————. 1957. Alpine lichens of western United States and adjacent Canada. I. The macrolichens. *Bryologist* 60:177–272. [Keys and notes (but no illustrations) of lichens occurring above 10,000 ft elevation.]

———— and I. M. Brodo. 1966. Biosystematic studies of *Lecanora pallida* and some related lichens in the Americas. *Beihefte Nova Hedwigia* 12:1–59. [Includes species in California related to *Lecanora caesiorubella*.]

Lamb, I. M. 1977. A conspectus of the lichen genus *Stereocaulon* (Schreb.) Hoffm. *Journ. Hattori Bot. Lab.* 43:191–355.

————. 1978. Keys to the species of the lichen genus *Stereocaulon* (Schreb.) Hoffm. *Journ. Hattori Bot. Lab.* 44:209–250.

Lawrey, J. D. 1984. Biology of lichenized fungi. Praeger, New York. [College-level introduction to lichenology with emphasis on ecology and physiology.]

Llano, G. A. 1950. A monograph of the lichen family Umbilicariaceae in the western hemisphere. Smithsonian Institution, Washington, D.C. [Illustrated monograph.]

Ornduff, R. 1974. Introduction to California plant life. University of California Press, Berkeley. [Detailed reference on vegetation of California.]

Sheard, J. W. 1974. The genus *Dimelaena* in North America north of Mexico. *Bryologist* 77:128–141.

Sierk, H. A. 1964. The genus *Leptogium* in North America north of Mexico. *Bryologist* 67:245–317. [Monograph with keys and illustrations including California species.]

Tehler, A. 1983. The genera *Dirina* and *Roccellina* (Roccellaceae). *Opera Botanica* 70:1–86.

Thomson, J. W. 1950. The species of *Peltigera* of North America north of Mexico. *Amer. Midl. Nat.* 44:1–68. [Monographic treatment with keys and descriptions.]

———. 1963. The lichen genus *Physcia* in North America. *Beihefte Nova Hedwigia* 7:1–172. [Monographic treatment with keys and descriptions.]

———. 1967. The lichen genus *Cladonia* in North America. University of Toronto Press, Toronto. [Complete illustrated treatment of *Cladonia*.]

———. 1987. The lichen genera Catapyrenium and Placidiopsis in North America. *Bryologist* 90:27–39.

Tibell, L. 1975. The Caliciales of boreal North America. *Symbol. Bot. Upsal.* 21(2):1–128. [Includes many of the California species.]

———. 1976. The genus *Thelomma*. *Bot. Not.* 129:221–249.

Tucker, S. C. 1973. New records and comments on lichens of California. *Bryologist* 76:209–211.

——— and W. P. Jordan. 1979. A catalog of California lichens. *Wasmann Journ. Biol.* 36:1–105. [A listing of all literature reports of California lichens.]

Weber, W. A. 1963. Lichens of the Chiricahua Mountains, Arizona. *Univ. Colorado Stud.* 10:1–36. [Keys (but no illustrations or descriptions) for many lichens also occurring in California.]

———. 1967. A synopsis of the North American species of *Cyphelium*. *Bryologist* 70:197–203. [Keys and discussions emphasizing the California species.]

———. 1968. A taxonomic revision of *Acarospora*, subgenus *Xanthothallia*. *Lichenologist* 4:16–31. [A list of the yellow species of *Acarospora* including many in California.]

Wetmore, C. M. 1960. The lichen genus *Nephroma* in North and Middle America. *Michigan State Univ. Biol. Series* 1:372–452. [Monographic treatment including California species.]

———. 1967. Lichens of the Black Hills of South Dakota and Wyoming. *Michigan State Univ. Biol. Series* 3(4):211–464. [Excellent keys (but no illustrations) to all lichen groups, including many California species.]

———. 1970. The lichen family Heppiaceae in North America. *Ann. Missouri Bot. Gard.* 57:158–209.

Range Maps

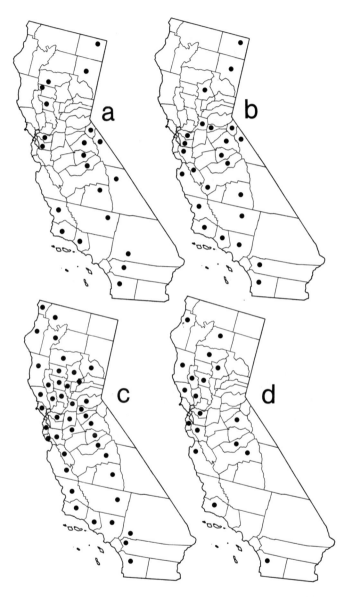

FIG. 72. Distribution of California lichens: (a) *Acarospora chlorophana* (b) *Caloplaca saxicola;* (c) *Candelaria concolor;* (d) *Collema furfuraceum.*

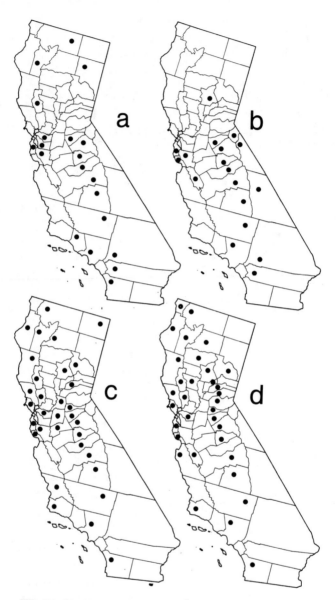

FIG. 73. Distribution of California lichens: (a) *Dermatocarpon miniatum;* (b) *Dimelaena oreina;* (c) *Diploschistes scruposus;* (d) *Evernia prunastri.*

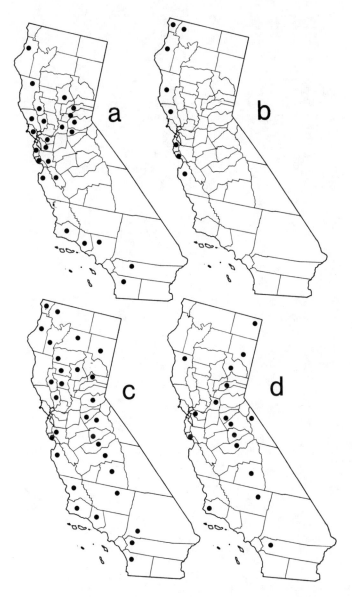

FIG. 74. Distribution of California lichens: (a) *Flavopunctelia fla-ventior*; (b) *Hypogymnia enteromorpha*; (c) *H. imshaugii*; (d) *Lecanora mellea*.

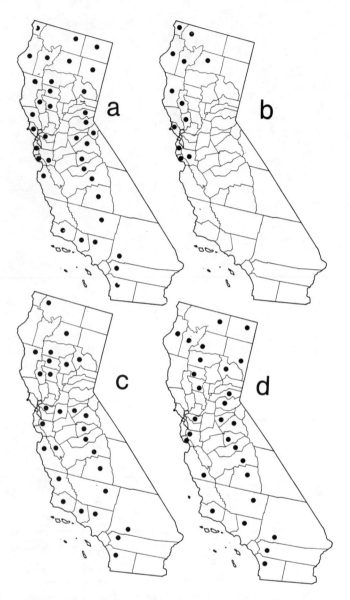

FIG. 75. Distribution of California lichens: (a) *Letharia vulpina;* (b) *Lobaria pulmonaria;* (c) *Melanelia glabra;* (d) *M. subolivacea.*

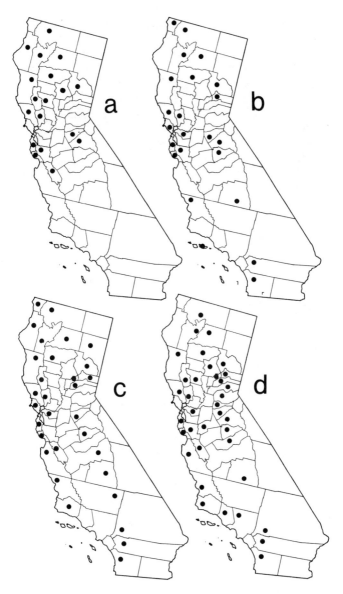

FIG. 76. Distribution of California lichens: (a) *Nephroma hel-veticum;* (b) *Parmelia saxatilis;* (c) *P. sulcata;* (d) *Parmelina quercina.*

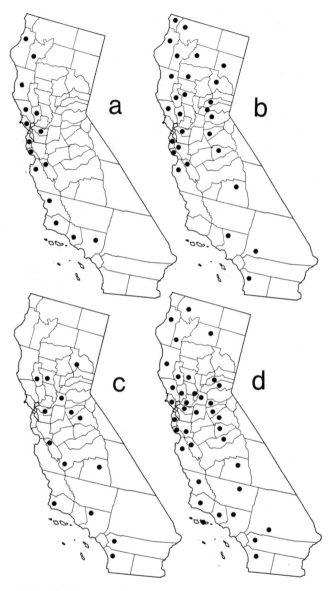

FIG. 77. Distribution of California lichens: (a) *Parmotrema chinense;* (b) *Peltigera collina;* (c) *Phaeophyscia orbicularis;* (d) *Physcia adscendens.*

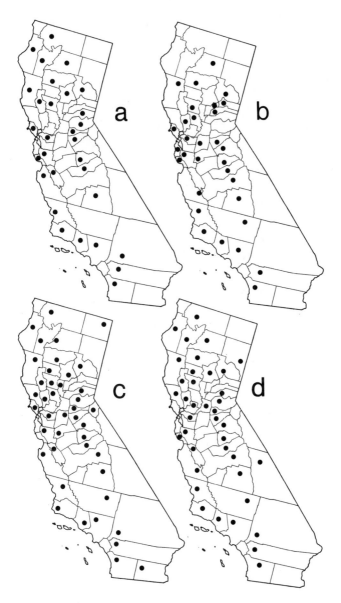

FIG. 78. Distribution of California lichens: (a) *Physcia aipolia;*
(b) *P. stellaris;* (c) *Physconia detersa;* (d) *P. distorta.*

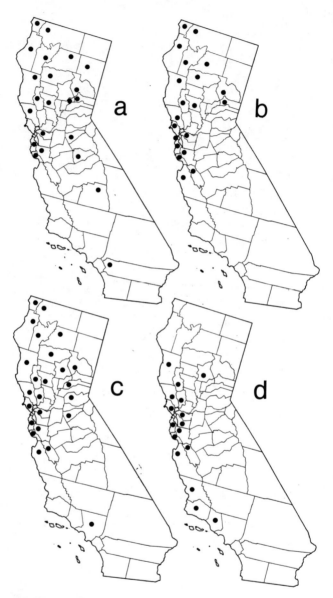

FIG. 79. Distribution of California lichens: (a) *Platismatia glauca;*
(b) *Pseudocyphellaria anomala;* (c) *P. anthraspis;* (d) *Punctelia
subrudecta.*

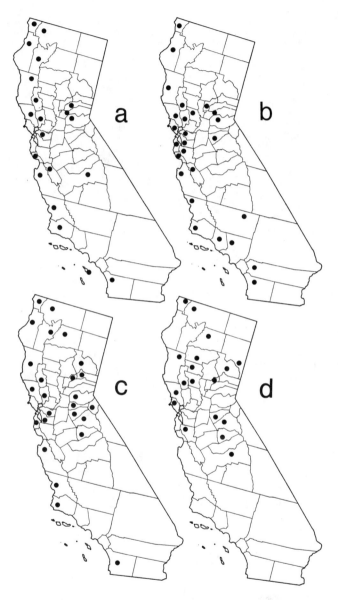

FIG. 80. Distribution of California lichens: (a) *Ramalina farinacea;* (b) *R. menziesii;* (c) *Tuckermannopsis canadensis;* (d) *T. platyphylla.*

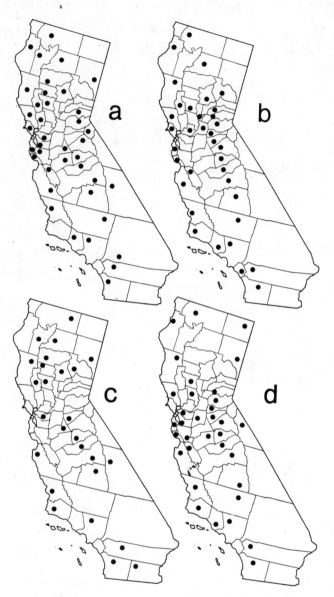

FIG. 81. Distribution of California lichens: (a) *Umbilicaria phaea;* (b) *Xanthoparmelia cumberlandia;* (c) *X. mexicana;* (d) *Xanthoria candelaria.*

Index

Designer: Rick Chafian
Compositor: G&S Typesetters, Inc.
Text: 10/12 Times Roman
Display: Helvetica
Printer: Consolidated Printers
Binder: Consolidated Printers &
             Mountain States Bindery

2/00   6   6/98
6/02   10   6/01
12/05   11   3/05
    11/10 (14) 9/10
  4/15 (16) 1/14